G000242030

Redshift

By Stuart Clark

University of Hertfordshire Press

First published 1997 in
Great Britain by
University of Hertfordshire Press
University of Hertfordshire
College Lane
Hatfield
Hertfordshire AL10 9AD

©Stuart Clark

ISBN 0 900458 66 6 Paperback
ISBN 0 900458 79 8 Hardback

Designed by
Beverley Stirling

Cover Illustration
Spiral galaxy M100 in the Virgo cluster of galaxies
(J Trauger and NASA)

Page layout by
Lisa Cordes

Printed by
Antony Rowe Ltd

Redshift By Stuart Clark

Dedicated to my mother and father, both of whom sadly passed away whilst I was writing this book.

I would also like to extend my heartfelt thanks to Nikk.

Contents

Mathematical notes

List of figures

List of plates

Preface

It is now seventy years since Edwin Hubble's work at Mount Wilson showed that the "starry nebulae" are in fact independent galaxies, most of them many millions of light-years away, and that the entire universe is expanding. Hubble's initial work depended on observations of the Cepheid variable stars, which 'give away' their real luminosities - and, hence, their distances - by the way in which they behave; but the revelation of the expanding universe depended entirely upon spectroscopic work. A receding object will be slightly 'reddened', and the amount of redshift in the spectral lines is a key to the rate at which the source is moving away. We still depend upon the redshift for our evaluations of distance beyond our immediate part of the universe, and its importance in modern astronomy cannot be over-estimated.

This new book, by Stuart Clark, is devoted entirely to the redshift phenomenon, but it is unusual in one respect. The author gives a clear, concise account of the redshift in all its aspects, but he also includes mathematical notes which are not found in ordinary 'introductions', and are usually confined purely to technical works which are quite beyond the scope of any but the trained specialist. What he has done, in effect, is to bring the two together, thereby making his book equally valuable to the beginner and to the qualified researcher.

He also includes a chapter about 'unconventional' interpretations of the redshift, and this is timely, because it has to be admitted that there are some lines of investigation which lead to contradictory results - so that there are suggestions that something is badly wrong somewhere! And it is undeniable that if there are serious errors in the conventional interpretation of redshift, it will be necessary to do some very serious re-thinking.

I read the book with great interest and enjoyment, and I am sure that many others will do the same. If you are interested in problems of modern cosmology - and most people are! - then this is the book for you.

Patrick Moore

Introduction

I can vividly remember standing at the University of Hertfordshire's observatory site (back in the days when it was still Hatfield Polytechnic) and looking up into the night sky. It was the early hours of the morning and I was alone amid the telescope domes. All was peaceful and tranquil, the silence broken only by the occasional cry of an animal in the surrounding woods and fields.

I had long since finished the work I had ventured out to do but somehow I could not drag myself away. As I gazed at the stars of the night sky and let my eyes wander along the Milky Way I felt truly at peace. It dawned on me that night that the universe is a wonderful playground for the mind. My imagination soared as I envisaged the destruction taking place at the black hole of Cygnus X-1, directly above me. I also tried to picture the internal structure of Antares and the nuclear fusion taking place within it. In my musing I was trying to conjure up images based upon my knowledge of physics, drawing inspiration from the starry dome around me.

It occurred to me that, just like any other playground, the universe too had rules to tell us what was acceptable and what was strictly forbidden. These rules are the laws of physics and some of the most fascinating to me have always been the ones which govern the behaviour of light. As human beings we rely upon light to give us a significant proportion of our sensory input. As astronomers we rely completely on light and the other, related electromagnetic radiations to provide us with an insight of the universe.

During my studies of astronomy, I have become particularly interested in the laws of light which are inextricably linked with those which govern the behaviour of the very fabric of the universe. The spacetime continuum, as this fabric is known, is in a constant state of expansion and causes one particular manifestation of the astronomical phenomenon, redshift. Amazingly, there are at least three different ways in which a redshift can be produced. Each of them is different from the other two. The Doppler Shift, the gravitational redshift and cosmological redshift will all be explored in this book.

In trying to explain the cosmological redshift, caused by the expansion of space, many texts refer to the easy to grasp but ultimately incorrect analogy of a Doppler shift. Whilst instantly gratifying to the reader, referring to it leads to problems when the concept is applied to the universe on its largest scales. This book explores much further and presents the three types of redshift in a deeper but ultimately more enriching fashion. I have also become intrigued by the possibility that there are other, more subtle ways of producing redshift which some astronomers have so far missed. I have explored those as well in a final, rather more speculative chapter.

In making this journey of discovery I have several people to thank. The first of these is Bill Forster of the University of Hertfordshire Press. Bill is a (sensibly) cautious man but in a weaker moment allowed me to talk him into publishing this book! I am really grateful for this because, instead of my book disappearing into the machinery of a large publisher and reappearing at the other end as something I no longer recognise, I have been allowed to help and assist with this book's passage into print. I have learnt an awful lot about the publishing world in the process. Two other people who contributed greatly to this book are the designer, Beverley Stirling and its type-setter, Lisa Cordes. Firstly, Beverley took every single one of the points on my design wish list and transformed them into something even better than I had imagined. I am particularly pleased with the front cover - the first draft of which spent some time on my office wall before I reluctantly gave it back! Secondly, Lisa took Beverley's basic designs and made them work for the actual text and figures, plates and notes. Although, this was no small feat, Lisa performed every task with good humour and has really triumphed. The layout of this book is impeccable.

The next person on this list of people who must have felt that Redshift had taken over their lives is my very own Nikki. Not only does she live with the cranky author but she made a considerable effort helping with the proof reading and rewriting. I strived to write this book in plain English with little or no unexplained technical jargon. That I succeeded, is due in no small part to Nikki. My grateful thanks are also extended to Dr. Chris Kitchin of the University of Hertfordshire's Observatory who read the manuscript from a technical point of view and made many helpful comments and suggestions. I am also indebted to Patrick Moore who provided the Foreword and made constructive criticisms to improve this book. Finally, the SCFC also deserve a mention: Don, Mick, Nick Sue, Christine and Chris for patiently waiting for this work to appear.

Stuart Clark
Welwyn Garden City. December, 1996

The nature of

light

The presence of light in the universe is what makes astronomy possible, since it is the analysis of this light, from distant celestial objects, which provides us with our knowledge of the cosmos. Light, however, is but one small part of a much larger whole: the electromagnetic spectrum. Radio waves, infrared and ultraviolet emission, x-rays and gamma rays all have a place within this distribution of radiation. Modern astronomers, with their arrays of sophisticated detectors, can now observe the universe in almost any part of the electromagnetic spectrum. Once collected, the electromagnetic radiation can be quantified in a number of different ways. The most obvious of these is to measure its intensity and deduce the brightness of the emitting object. To an astronomer, one of the radiation's most important quantities to be measured is its redshift.

In order to understand the redshift however, the nature of light itself must first be understood. This has not proved an easy task and, in fact, it has puzzled mankind for nearly three millennia. The classical Greek philosophers such as Plato, Aristotle and Pythagoras all pondered the nature of light. The first correct, fundamental assumption about light was made by Hero of Alexandria in the fourth century. He stated that light travels along the shortest path between two objects. The work of the ancients continued until the Roman Empire fell in the fifth century and the world lapsed into the Dark Ages. Following the end of these scientifically desolate times, European scientists began to tackle the problem in the latter half of the thirteenth century. To begin with progress was slow and the next four hundred years or so were characterised by the gradual development of lenses and mirrors. Little thought went into the actual, physical nature of light.

That work began in earnest during the seventeenth century and has evolved into our modern view which is that no single theory can account for the behaviour of light. Instead, two apparently contradictory schools of thought have been developed and

Mathematical note 1.1

Wavelength and frequency of light

When multiplied together, the wavelength and frequency of a wave give its speed of propagation:

$$v = f\lambda$$

where
v = speed of wave propagation
f = frequency
λ = wavelength

In the case of electromagnetic radiation, the speed of the wave's propagation is always a constant:

$$c = f\lambda$$

where
c = speed of light in a vacuum

Hence:

$$f \propto \frac{1}{\lambda}$$

where
$1/\lambda$ is called the wave number

reconciled. One is known as the electromagnetic theory whilst the other is termed the quantum theory. Instead of thinking of the two as contradictory, it would be better to view them as complementary. The same, yet different. Counterparts, in fact, since the two theories make up our understanding of light in the same way as the counterparts of man and woman make up mankind. When thinking of the interaction of one ray of light with another, the electromagnetic theory must be used. When light is interacting with matter, the quantum theory comes into play.

Light as a wave

The electromagnetic theory states that light is a wave and as such, has the properties of wavelength and frequency. Wavelength is the distance between successive peaks or troughs whilst frequency is the number of peaks or troughs which pass a specific point in space every second. Wavelength and frequency are interrelated quantities (see *mathematical note 1.1*). Although thought of as a wave, light is atypical in some respects as will become apparent.

In general, waves come in two forms, longitudinal or transverse (see *figure 1.1*) and, whatever the type, are transmitted through a medium. Waves, themselves, are not objects which travel through a medium but, instead, are disturbances within a medium.

A medium is anything through which a wave can travel. The medium can be a solid, a liquid or a gas. Because the medium is not moved, as a whole, in the direction of the wave's motion, the wave is said to propagate through the medium.

The individual particles of matter which make up the medium do not actually move with the wave. Instead, they are disturbed by its passage before returning to their original positions. Whilst being disturbed, they communicate this fact to their neighbouring

Figure 1.1
Transverse waves

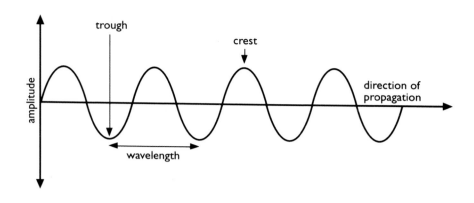

Longitudinal waves

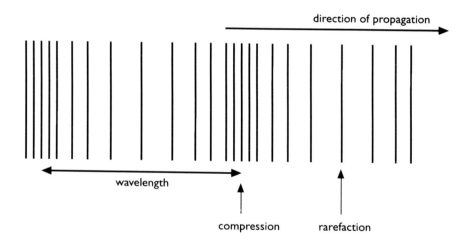

particles. They too then suffer the disturbance, allowing the first particle to return to its original position. In this way the wave is propagated through the medium. The definition of the type of wave is based upon how the particles in the medium move in order to propagate the wave.

A longitudinal wave is one in which the medium is compressed in the direction of the wave's movement. Increasing the density of matter at one point in the medium, leads to a decrease in its density just behind the increase. The increase is known as a compression whilst the decrease is termed a rarefaction. An example of this is a sound wave. If you look carefully at one of the loud speakers of a stereo music system, you can observe the speaker's diaphragm moving in and out in time with the music. This movement is causing the air to be alternately compressed and stretched, depending upon whether the diaphragm is moving in or out. This sets up a longitudinal wave which carries the sound of music to our ears.

In a transverse wave, the disturbance is at right angles to the direction in which the wave travels. The size of that disturbance is known as the amplitude of the wave. Particles in the propagating medium are disturbed sideways and oscillate about a fixed position. Only after a particle has started to move can the one in front of it begin its oscillatory motion. In this way a time lag is introduced between the motion of adjacent particles. If their positions at any instant in time are drawn on a graph, a wavy line, known in mathematical parlance as a sine wave, is produced. An excellent example of a transverse wave is the moving disturbance in a piece of string caused by the flick of a wrist. At no time does the string actually move away from you but the 'hump', created by the flick of your wrist, moves down the string like a water wave approaching the sea shore.

During the seventeenth century all sorts of phenomena, associated with the way in which rays of light move and interact with one another, were being observed. Obviously, attempts were being made to explain what was being seen and, in an effort to understand some observations of what happens to light when it passes through a thin film, Robert Hooke proposed the idea that light was a rapid vibration of the medium through which it passed. This laid the foundations for the wave theory of light.

The wave nature of light was by no means universally accepted in the seventeenth and eighteenth century, despite great work by scientists such as Hooke and Christiaan Huygens of Holland. Although Huygens used a wave theory of light to determine correctly the laws of refraction and reflection, a rival theory, known as the corpuscular theory, remained very popular. This stated that light was a stream of particles emitted from any bright body. Its greatest advocate was none other than Sir Isaac Newton, whose body of work otherwise greatly enhanced our understanding of gravitation and optics. Even though the evidence for the wave theory grew, the weight of Newton's name hung like a protective blanket over the corpuscular theory, sometimes known as the emission theory, with the result that little progress was made during the eighteenth century.

In 1801 an Englishman, Thomas Young, began to present a series of papers which sought to prove the wave theory. Conscious of the fact that Newton's name still made the corpuscular theory the one of choice, he was at pains to point out that his work was based on Newton's research. Nevertheless, because he was a wave advocate, his work was treated harshly and Young became despondent at the response. Just over a decade later, a French scientist, Jean Fresnel, began to build on the work of Christiaan Huygens, arriving at many of the same conclusions that Young had done. Whilst performing these calculations, Fresnel was unaware of Young's previous work. When the two finally compared notes they pooled their resources and became joint advocates of the wave theory.

In the middle of the nineteenth century, the final pieces of the puzzle fell into place. The first successful terrestrial measurement of the speed of light was performed by Armand Hippolyte Louis Fizeau in the suburbs of Paris. Across the Channel, Michael Faraday was working on the separate problem of electromagnetism. In trying to elucidate its nature, Faraday established an interrelationship between light and electromagnetism. He did this by showing that the properties of light could be changed by a strong magnetic field. One of his peers, James Clerk Maxwell, made the necessary intellectual leaps in order to formulate succinct mathematical laws for electromagnetism shortly afterwards. Using those laws, Maxwell showed, in 1864, that an electromagnetic field could propagate through a medium as a transverse wave. In deriving an expression for the speed at which this wave propagated, Maxwell theoretically derived a value which was too close to Fizeau's measured speed of light to be a coincidence! The nature of light was obvious, concluded Maxwell: it was none other than an electromagnetic wave.

Light and its associated radiations, became known as electromagnetic radiation. It is a kind of transverse wave and the disturbances which mark its passage are oscillations in electric and magnetic fields. These oscillations take place at right angles to each other and also to the wave's direction of propagation (see figure 1.2).

Figure 1.2
Electromagnetic radiation

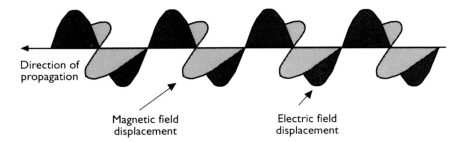

Direction of propagation

Magnetic field displacement

Electric field displacement

According to the electromagnetic theory, the differences between the various types of radiation are due to the differences in their wavelength and frequency. The shortest wavelengths are possessed by gamma radiations whilst the radiation with the longest wavelengths are the radio waves (see figure 1.3).

Despite these great advances, however, one erroneous assumption was perpetuated. This was that, since light was a transverse wave, it required a medium in which to propagate.

Figure 1.3
The electromagnetic spectrum

The numerical quantities of wavelength, frequency and photon energy are listed here for the boundaries between different regions of the electromagnetic spectrum. For example, infrared radiation is any electromagnetic wave which has a wavelength between 7.0×10^{-7} metres and 1×10^{-3} metres in length. Radio wares are electromagnetic radiations with wavelengths greater than 1.0×10^{2} metres

Wavelength	Frequency	Photon energy
	Gamma rays	
1×10^{-11} metres	3×10^{19} hertz	1.99×10^{-14} joules
	X-rays	
1×10^{-8} metres	3×10^{16} hertz	1.99×10^{-17} joules
	Ultraviolet	
3.5×10^{-7} metres	8.57×10^{14} hertz	5.68×10^{-19} joules
	Visible light	
7.0×10^{-7} metres	4.29×10^{14} hertz	2.84×10^{-19} joules
	Infrared	
1.0×10^{3} metres	3.00×10^{11} hertz	1.99×10^{22} joules
	Microwaves	
1.0×10^{2} metres	3.00×10^{10} hertz	1.99×10^{23} joules
	Radio waves	

The medium was termed the luminiferous aether. Following the proof of the wave nature of light, many scientists began searching in earnest for the aether. No one was successful and it was not until the publication of Einstein's special theory of relativity that the reason for this was made clear, as will be seen.

The speed of light

The senses possessed by humans are not sharp enough to detect naturally the time it takes for rays of light to emanate from a light bulb, reflect off objects contained within a room and finally strike our eyes. There appears to be no delay in flicking a switch and seeing the room illuminated before us. In fact, it appears as if light travels instantaneously from one place to another. Although this particular point has been illustrated with a twentieth century example, humans have always wondered at the apparently instantaneous nature of light. For millennia it was thought that light indeed travelled with infinite speed. No sooner had it left one place than it had arrived at another!

The work which was performed on light during the middle ages led some scientists to the sneaking suspicion that light was not, in fact, instantaneous but did have a finite speed. The speed was, obviously, exceptionally high but it was not infinite.

The first scientist to act on these suspicions was Galileo. He attempted to measure the speed of light by taking two lanterns out on a dark night. He positioned himself on a hill whilst an assistant took the second lantern and stood on another hill three miles away. Both lanterns were then covered with buckets and the experiment begun. Galileo would lift the bucket and allow his light to shine. At the same time he would begin the timing. When his co-worker saw Galileo's light, he would lift the bucket covering his light. Galileo would then stop the timing when he saw the light of his assistant's lantern. Since Galileo knew the distance between the hill tops, he reasoned that the time interval would allow the speed of light to be calculated. His results were inconclusive, however, because they kept giving very different figures for the speed of light. What Galileo had not realised was that the speed of his and his co-worker's reactions were too slow compared to the speed of light. The time it took them to react to one another was far too long, compared with the time it took for the light ray to cross the three miles. In fact, we now know that in the time in took them to react, light could have travelled the distance between them tens of thousands of times.

Ironically, in 1675, forty three years after the death of Galileo, it was one of the great man's discoveries which led to the first successful measurement of the speed of light. Six and a half decades previously, Galileo had discovered four moons in orbit around Jupiter, the fifth planet in the solar system. In 1675, the Danish astronomer, Ole Christensen Römer, measured the time it took for each of these moons to orbit Jupiter, a quantity known for each as its period. To his surprise, repeat experiments gave different answers. In an attempt to

Figure 1.4
Romer's observation

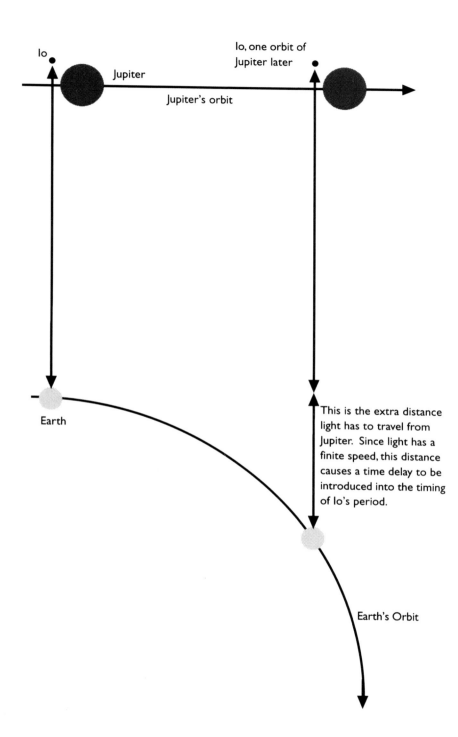

Io

Io, one orbit of
Jupiter later

Jupiter

Jupiter's orbit

Earth

This is the extra distance
light has to travel from
Jupiter. Since light has a
finite speed, this distance
causes a time delay to be
introduced into the timing
of Io's period.

Earth's Orbit

understand why this was so, Römer began to think about the positions of the Earth and Jupiter in their orbits when the observations were made. At the beginning of the timings, the Earth would be at one distance from Jupiter but in the time it took for a Jovian moon to complete one orbit, the Earth would have moved along in its orbit and so the distance between it and Jupiter would be different (see figure 1.4). If that distance had lengthened, i.e. the Earth were moving away from Jupiter, the period of the moon in question would be longer than if the Earth were moving towards Jupiter when the timing was made.

If light travelled instantaneously, these differences in distance would not affect the orbital period of the moon. To explain his observations Römer concluded that light did travel with a finite speed. That being the case, his measurements included the time in which light travelled the difference in the distances between Earth and Jupiter at the beginning and end of each observed period.

Following this important step, which highlighted a phenomenon which could only take place if the speed of light were a finite quantity, a corroborating phenomenon was needed. Quite by accident, such evidence was forthcoming in 1727 when the Englishman, James Bradley (who was the third Astronomer Royal), discovered the phenomenon he went on to call aberration. Bradley was searching for a method with which to measure the distances to nearby stars. He reasoned that a nearby star would appear in slightly different locations, relative to the background stars, when viewed from Earth at six monthly intervals. This apparent change of position is caused by our viewing the star from different positions in the Earth's orbit. It is a phenomenon which has come to be known as parallax.

Parallax is directly analogous to holding a finger in front of your face, at a distance of about eighteen inches and looking at it with alternate eyes, whilst keeping the other shut. If this is done, you will notice that your finger appears in a different place, relative to the background objects, depending upon which eye you are looking through. In this analogy, your individual eyes are lik xe the Earth at different points in its orbit. Your finger represents the nearby star and the background objects represent the distant stars.

When Bradley performed these observations, he did indeed detect a shift in the position of the star at which he was looking. Upon analysis, however, it was too large a difference to be due just to parallax. The star he measured was Gamma Draconis and, over the period of six months, he found the difference in direction to be about forty seconds of arc. [One second of arc is a sixtieth of an arcminute which is itself a sixtieth of a degree. This means that an arc second is exactly one three thousand, six hundredth of a degree ($1/3600^0$)] Further investigation showed that all objects in the night sky shifted back and forth about an average position by this same amount. Bradley called this unexpected phenomenon 'aberration' and set about explaining how it arises.

The reason is that, even though the telescope is pointing directly at the star in question, the image appears off centre, because the telescope is carried along slightly (by the Earth moving through its orbit) while the light is travelling down the tube. In order for the star to be centred, the telescope has to be tilted a little. Thus the telescope is not pointing precisely in the actual direction of the star. Since it appears in the centre of the telescope's field of view, however, that is the position to which it is ascribed. As the Earth moves in a circular orbit, six months later it is travelling in the opposite direction and the star appears displaced in the opposite direction, too. Thus the telescope has to be tilted in the opposite direction and the star appears to have changed position.

The angle at which the telescope is pointed allows the speed of

Mathematical note 1.2

Calculating the speed of light from aberration

Aberration causes an annual displacement in a star's position of 41 seconds of arc. This means that the maximum displacement, at any one time, from the star's true position is 20.5 arc seconds. This maximum displacement is caused when the light from the star arrives at right angles to the motion of the Earth and is in the direction of the Earth's motion. Constructing a vector diagram to show this gives:

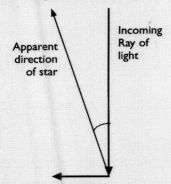

Motion of the Earth (30,000 m/s)

where ϑ = maximum displacement

This diagram is actually nothing more than a right-angled triangle. Hence, by treating the motion of the Earth vector as the opposite and the incoming light ray vector as the adjacent, simply trigonometry gives:

adjacent = opposite
           Tan J

By assigning the speed of the Earth's motion through space to the opposite and the maximum displacement to J, this equation yields the speed of light as the adjacent.

light to be calculated, providing that the speed of the Earth through space is known (see mathematical note 1.2). Although the value which Bradley calculated was none too accurate by today's standards, his analysis of the phenomenon allowed those who believed in a finite speed of light to enhance significantly their theories. Those scientists who clung to the opinion that light propagated instantaneously were left floundering by Bradley's result.

It was to take over a century before engineering skills were sufficiently developed to allow an experiment to be constructed which could test the theories about the speed of light. In 1849, Fizeau updated the method tried, unsuccessfully, by Galileo over two centuries earlier.

In the suburbs of Paris, Fizeau set up a mirror and a candle five miles from one another. It was his intention to measure the time it took the light to travel the ten mile round trip. In front of the candle, Fizeau placed a cogged wheel which was attached to a weighted pulley system. The candle was lit and the pulley set in motion so that the cog-wheel turned in front of the candle. This caused the light emitted by the candle to be 'chopped-up' and sent off in pulses. When this light reflected off the mirror, it began its journey back towards the cogged wheel. Fizeau observed the mirror through the cogs so that, at times, his view was blocked, whilst at others, he could see the mirror clearly. If the returning pulse of light arrived at the wheel when there was a gap rather than a cog, Fizeau would see a flash. He then instantly knew that the rotation of the wheel was synchronised with the time it took for the light to travel the ten mile round trip. The weight on the pulley could be varied so that the rotational speed of the cogged wheel could be adjusted. If Fizeau increased the velocity of the wheel so that the returning pulse passed through the adjacent gap to the one through which it had escaped, he also knew how far the wheel had turned. Combining this measurement with the angular

Mathematical note 1.3

Fizeau's method of calculating the speed of light

The distance through which the light was travelling was known by Fizeau to be ten miles. All he needed was a time measurement in order to complete his calculations.

$$v = \frac{s}{t}$$

where
v = speed
s = distance
t = time

Fizeau could get his time measurement by using the cogged wheel. He had to know three things:

α = total number of cogs on the wheel
β = angular speed at which the wheel turned
γ = number of cogs which pass him before the light ray returns

All three variables are relatively easy to measure and can give the time taken for the light ray to complete its round trip if they are combined as follows:

$$t = \frac{\gamma}{\alpha\beta}$$

Substituting into the original equation gives:

$$c = \frac{s\alpha\beta}{\gamma}$$

where
c = the speed of light

speed of the wheel meant that Fizeau could calculate the time it took for light to travel the ten mile round trip and hence work out the speed of light (see *mathematical note 1.3*).

In this way, Fizeau made his determination of the speed of light. Although that figure turned out to be a little too high, considering the era in which he made his measurements, it was a remarkable result.

The definitive measurement of the speed of light was made in 1926 by the American physicist Albert Abraham Michelson. In a similar type of experiment to Fizeau's, Michelson set up a mirror to reflect a source of light. The separation between the source and the mirror for this experiment was twenty two miles. The light leaving the source struck a nearby mirror, which reflected it in the direction of the distant mirror. The nearby mirror was a rotating hexagon, which continued to turn whilst the light covered its forty four mile round trip. If the rotation of the mirror was synchronised precisely, the returning beam of light would strike a mirror panel in exactly the same position as the one it initially left. This would direct the light back to Michelson. Using a similar mathematical approach to Fizeau's, the number of turns of the hexagon and its angular speed could then be used to calculate the speed of light.

Today, the accepted value of the speed of light is 2.99792458×10^8 metres per second (m/s). Although this is termed the speed of light it is actually the speed at which all electromagnetic radiation propagates through space. When travelling through denser media such as water, glass or even air, the speed is slower. Through air, however, the difference is virtually negligible. The precise speed depends upon the optical properties of the medium. For simplicity, this book will refer to the speed of light (in space) as being 3×10^8 m/s.

The luminiferous aether

With the demonstration during the nineteenth century that light was, indeed, a wave, the search was on to find the postulated medium through which it propagated. Since all other known wave phenomena required a medium through which to move, it was reasoned that one for light must also exist. It was known, however, that space was a vacuum so the medium could not be air. The name given to light's medium was the luminiferous aether, often shortened to just aether or ether.

It was thought that certain, specific investigations of light would allow the properties of the all pervading ether to be uncovered. Scientists were quick to realise that a welcome by-product of the discovery of the ether would be to provide, finally, a framework within which all motion in the universe could be referenced. Needless to say, the possibility of finding something which could be used to measure the universe in an absolutely precise way was very appealing. The concept became known as absolute space and many scientists began searching for it.

In the same way that a sailing ship cuts through water in the ocean, so the Earth was thought to cut through the ether. Every other object in the universe was also thought to do the same. Since the motion of a sailing ship can be detected by measuring the flow of water past the hull, physicists of the nineteenth century tried to conceive of an experiment which would detect the flow of the ether: the ether wind.

They reasoned that, if the Earth travelled through the ether, the medium through which light propagated at 3×10^8 m/s, then one effect should be obvious: stars should change their focal positions in telescopes every six months. The reasoning behind this was that at one point during the Earth's orbit, our planet is travelling towards the star in question whilst six months later it is moving away. Thus, when the Earth was travelling towards the star, the light would enter the telescope and begin travelling

Figure 1.5

The expected ethereal change of focus

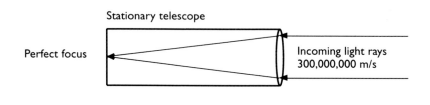

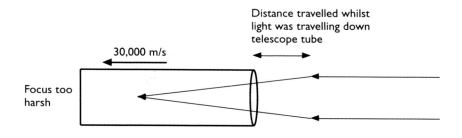

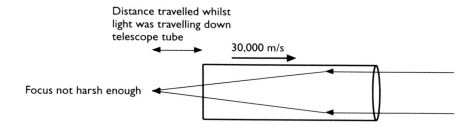

down the tube. Once the telescope was moving towards the star, the base of the telescope, containing the eyepiece, would be moving towards the light ray. The two would meet before the light ray had focused sufficiently. The eye piece would then need to be adjusted to compensate for the motion of the telescope. Six months later, however, when the telescope was moving away from the light ray, that focus would have to be relaxed in order for the image to be sharp when it finally caught up with the eye piece (see figure 1.5). The effect was searched for and not found. Only with the introduction of Einstein's special theory of relativity could this be explained satisfactorily. At the time, however, the null result was explained away by Fresnel who contrived an idea that the ether was denser in certain places and was dragged along by bodies moving through it.

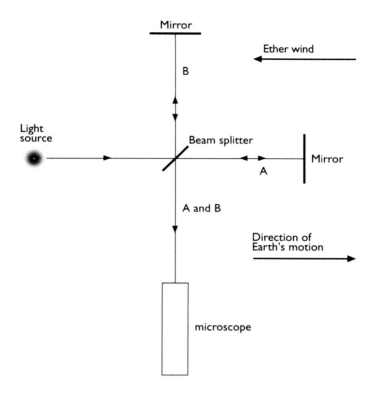

Figure 1.6
The Michelson-Morley interferometer

The crucial experiment in the discussions of the ether was performed in 1881 by Albert Abraham Michelson and Edward Williams Morley, forty seven years before Morley's precise determination of the speed of light. In the Michelson-Morley experiment, a beam of light is split into two and sent in different directions along identical length journeys from one another. Following their return , they are combined and any difference between them is checked for *(see figure 1.6)*.

Since the experiment was performed on the Earth, which is in motion through space and hence the hypothetical ether too, one ray of light was dispatched along the Earth's direction of travel, whilst the other was directed at ninety degrees away from this. It was expected that the ray in the direction of travel would have to push itself through the ether wind head on, whilst the ray at ninety degrees would only have to deal with an ethereal 'cross-wind'. This should affect the passage of the light so that the two rays, despite travelling the same distance, would not travel it in the same period of time. In fact, when the experiment was performed, the time taken for both rays to travel along their paths was identical. The conclusion was inescapable: based upon the Michelson-Morley experiment, it is impossible to detect the Earth's motion through space. In fact, the results achieved were the ones which had been expected if the ether did not exist!

The trouble was that the belief that the ether was a real thing was so firmly entrenched that the scientists of the day were unwilling to give it up easily. The mental cartwheels which many went through in order to rationalise Michelson and Morley's results gave rise to four theories which tried to explain why the ether had not been detected.

The first explanation was that the Earth was stationary in space and so not moving through the ether. This was rejected without ever being seriously considered since the evidence that the Earth moved was overwhelming.

The second possibility was that the Earth dragged the ether along with it, which would prevent the light rays being affected by the ether wind. The objection to this idea came from Bradley's observations of aberration. The key reasoning behind understanding that phenomena was that the Earth moved through the ether and the light moved in straight lines in the ether. Thus another possibility was rejected.

The third idea was that the speed of light was a constant with respect to whatever had emitted it. This meant that the light rays would both be travelling at 3×10^8 m/s, regardless of how fast the interferometer was moving. This was an idea which would also explain the failure of astronomers to detect a shift in the position of focus for stars. Although difficult for scientists of the day to accept, this idea became central to the development of Einstein's theory of special relativity.

The fourth theory proposed to explain the results of the Michelson-Morley experiment became known as the Fitzgerald-Lorentz contraction. It was an ad hoc hypothesis, dreamt up in order to explain the experiment's failure to detect the ether. It relied on a notion that perhaps objects were squashed as they travelled through the ether. It was a tenuous argument, though, which did not take into account factors such as the physical strength of objects. Nevertheless, despite the reasoning behind it being wrong, Einstein proved in his special theory of relativity that a contraction in the direction of motion is indeed observed to occur.

The special theory of relativity

Looking at the work of scientists before him, Einstein concluded two things: firstly, that the ether could not be detected and secondly, that the speed of light was a constant. These became known as the postulates of special relativity and are statements of fact which rest solidly upon observation. The first postulate was probably Einstein's, scientifically, most diplomatic move. He stated that the ether could not be detected but did not declare outright that it did not exist! Anyone who still wished to believe in the existence of the ether could do so but, from Einstein's point of view, it could be discounted. This robbed him of an absolute space against which velocities could be measured. Everything now had to be measured relative to something else. For instance, moving objects on the Earth have their velocities measured relative to the Earth. If we were to measure the velocity of a car relative to the Sun, then the velocity of the Earth would have to be added to that of the car.

An immediate consequence of the first postulate is that without anything to measure yourself against, it would be impossible for you to know if you were moving with a constant velocity or if you were stationary. This was illustrated by the Michelson-Morley experiment which failed to detect the motion of the Earth.

The second postulate flies in the face of common sense but is, nevertheless, as true as the fact that the Earth is round! Everyone is used to the concept that a ball thrown from a moving vehicle will have the speed of the vehicle superimposed upon the speed with which the ball is thrown. If the ball is thrown in the direction in which the vehicle is travelling, the total speed of the ball will be the speed at which it is thrown plus the speed at which the vehicle is moving. If the ball is thrown backwards, the speeds will have to be subtracted.

The phenomenon of aberration tells us that this is not true of light. Aberration is a constant, regardless of which star is considered. Every single star is offset by 20.5 arcseconds, regardless of how that star is moving through space. Some will, doubtlessly, be moving towards us whilst others will be moving away from us, yet all of them will emit light which travels to the Earth at 3×10^8m/s. If this speed were not true and the individual speeds of the stars had to be added to or subtracted from the speed of the emitted light, the aberration offsets would be different for each star.

The fact that a change of focus could not be detected when scientists were searching for the ether is another confirmation of the constant speed of light. Since the light rays always travelled down the telescope's tube at 3×10^8m/s, regardless of the direction in which the telescope was moving, they always came to focus at the same position.

Using these two postulates, Einstein began to make deductions about real-life situations which they would influence. He chose to examine objects, which were moving at a constant speed and work out how they would appear to people observing them. He discovered that the way they appeared would be affected by the properties of light, especially its finite velocity.

Before proceeding further, it is worth noting that one of the barriers to understanding relativity is that it is awash with terminology. In fact, most of the terminology is easy to grasp. With the dismissal of the ether, the universe no longer has a framework against which velocities can be measured. It is therefore necessary to constantly state which frame of reference is being used. For example, a moving object carries with it a frame of reference which is different from a stationary object. A frame of reference can be most easily thought of as being a set of co-ordinate axes and a clock. These allow observers to locate objects in both space and time. An event is an observed happening for which a location in space and the time at which the happening took place are known. An observer is anyone or anything capable of making an observation. Every observer has a frame of reference depending upon his/her/its state of motion. Often the word inertial is used to preface the terms, frame of reference and observer. The word refers to the fact that these things must be either stationary or travelling at a constant speed with respect to one another. This is because special relativity does not deal with objects which are changing their velocity, i.e. accelerating or decelerating. The word inertial accentuates this.

To highlight the way in which one's state of motion affects one's perception of the universe, imagine two cars parked on a road. Initially, they are side by side and their clocks are in perfect synchronisation. One of the cars begins moving away from the other and reaches a constant velocity. At some arbitrary distance, someone in the moving car flashes a torch back towards the stationary one. A set amount of time later, the same person flashes the torch again. The observer in the stationary car and the sender are both measuring the time between torch flashes. Upon comparison, they discover that the time between pulses is greater for the receiving observer than the time between pulses measured by the sender.

This is because the second car is moving away from the first at a constant velocity. Hence, by the time the torch is flashed a second time, the distance between the two cars has increased.

Since the speed of light is finite it takes a longer period of time to cross the increased distance between the two cars (see *figure 1.7*). Hence, the stationary car measures the time between the pulses as being longer than the moving car measures it to be (see *mathematical note 1.4*). It is a similar kind of observation to that made by Römer when studying the changing period of Jupiter's moon, Io.

Just from the simple example using the cars, it can be seen that the way an observer perceives the objects and events within the universe is dependent upon his/her or its state of motion. Had the two cars been stationary or moving together with the same velocity, no time difference between the torch flashes would have been noticed. Perceptions are progressively warped the closer the speed of an object gets to the speed of light. Fast moving objects must be identified so that the effects of relativity can be taken into account.

Splitting light into a spectrum

In tandem with the work into light's propagation, which culminated in the theories of relativity, work investigating light's properties of wavelength and frequency were continuing. One of Newton's greatest pieces of work was realising the phenomenon of white light being split into a rainbow of colours. This separation of white light into a spectrum can be easily achieved by passing the beam through a glass prism. As a result of his experiments, Newton drew the correct conclusion that white light was composed of all the colours mixed together.

In 1814, whilst performing this experiment with a beam of sunlight, German optician Joseph Fraunhofer magnified the rainbow spectrum and made a surprising discovery. In his magnified spectrum he saw a multitude of fine dark lines, counting over six hundred of them. Fraunhofer had observed what became known as an absorption spectrum.

Figure 1.7
**Two cars measuring
light pulses**

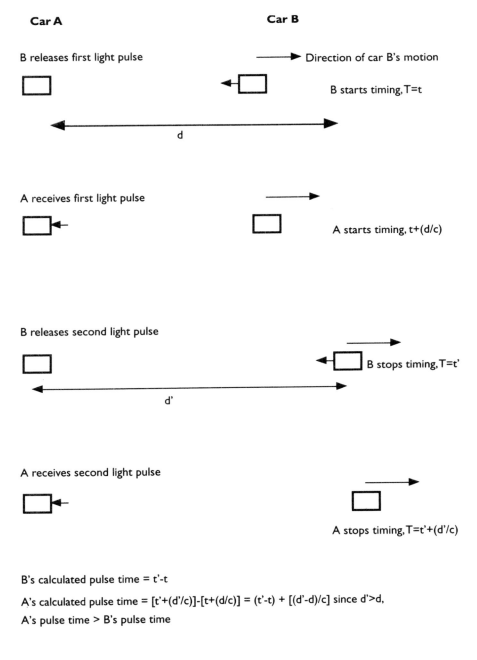

Car A   **Car B**

B releases first light pulse   Direction of car B's motion

B starts timing, T=t

d

A receives first light pulse

A starts timing, t+(d/c)

B releases second light pulse

B stops timing, T=t'

d'

A receives second light pulse

A stops timing, T=t'+(d'/c)

B's calculated pulse time = t'-t

A's calculated pulse time = [t'+(d'/c)]-[t+(d/c)] = (t'-t) + [(d'-d)/c] since d'>d,

A's pulse time > B's pulse time

Nearly fifty years later its counterpart, an emission spectrum, was observed. German chemist, Robert Bunsen had invented a special gas burner which produced a largely colourless flame. It was designed to be used to flame test chemical elements.

Mathematical note 1.4

Calculation of the k-factor

The precise increase in the two cars' timing is governed by something known as the K factor. In order to derive the mathematical expression for the K factor, plotting the two cars on a spacetime diagram will be useful. A spacetime diagram is one in which the vertical axis denotes time and the horizontal one denotes space: a distance perhaps. It makes no difference in which direction this distance occurs A stationary object would be represented by a vertical line whilst a moving object would cause its line to tip one way or the other. The lines themselves are known as world lines. Since special relativity can only deal with objects which are either in constant uniform motion or stationary, the lines are always straight. The shallower the gradient of the line, the faster the object is travelling. A curved line would indicate that either acceleration or deceleration was taking place. Light rays, which travel at the speed of light are represented on spacetime diagrams as diagonal lines at forty five degrees to each axis. In the case of the two cars example, the spacetime diagram would look like this:

The diagram has been constructed so that it shows everything from car A's frame of reference. This is obvious by the fact that A's world line is parallel to the time axis and perpendicular to the space axis. The time between flashes of the torch, as measured by each car, is represented by the distance along their world lines between the two light rays. This is longer for car A than for car B. In fact, the time between flashes from car B has to be multiplied by a factor, K, in order to give the time which A measures between the flashes. The faster B travels away from A, the greater K must be. In fact, K is entirely dependent upon the relative velocity of separation between A and B. Note, however, that without proper units on the axes these cannot be regarded as being to scale.

The best way to measure this velocity is for car A to time a flash of light which it bounces off car B. In this example, relativistic units will be used which means that the speed of light is treated as if it were 1.

The spacetime diagram of this situation looks like this:

The diagram shows the following things. The two cars begin at the same point. Car B moves away from car A and after a time, T, car A sends a pulse of light towards car B. Because they are moving apart with a constant velocity, the light pulse arrives at car B after a time, KT. It is reflected back towards car A. Since the reflection is as if car B were to emit a signal after time, KT, car A receives the reflected pulse after a time K^2T from the start of the experiment. The time taken for the light signal to reach car B is given by the equation:

$$t = T + \frac{1}{2}\left(K^2T - T\right) = \frac{1}{2}\left(K^2 + 1\right)T$$

The position of car B can be easily calculated with the equation:

$$x = \frac{1}{2}\left(K^2 - 1\right)T$$

Note: this equation should then be multiplied by the speed of light, c, but since we are working in relativistic units the speed of light is taken as 1.

Using these two equations the velocity can be easily calculated:

$$v = \frac{x}{t} = \frac{\frac{1}{2}\left(K^2 - 1\right)T}{\frac{1}{2}\left(K^2 + 1\right)T} = \frac{\left(K^2 - 1\right)}{\left(K^2 + 1\right)}$$

Rearranging for K gives:

$$K = \sqrt{\left(\frac{1 + v}{1 - v}\right)}$$

Note: In non-relativistic units, each 1 would be replaced by c.

time | A | Light pulse's world line
KT
B
T

time | A | B
K^2T
T
KT
space, x

Bunsen's colleague, Gustav Kirchoff suggested that they study the resultant flames by passing the light through a prism. Together they built their spectroscope and discovered, in short order, that each chemical element tested displayed a characteristic pattern of spectral lines. Spectra, it was realised, were the atomic equivalent of human fingerprints!

By now, common scientific wisdom recognised that the various colours of light represented different wavelengths of electromagnetic radiation. It was also understood that the electromagnetic spectrum continued on either side of the visible section. Longer wavelengths than light were found after the red end of the spectrum and shorter wavelengths were found before the blue. By studying light passing through a cloud of gas, absorption and emission spectra can be seen depending upon the angle at which the observer is looking (see *figure 1.8*). Studying chemical elements in this way, it became obvious that the wavelength of the absorption lines were exactly the same as the wavelength of the emission lines. In some way, the atoms could absorb and emit only those wavelengths which were characteristic of their chemical element. Why atoms should behave in this way was unanswerable by the wave theories of light. The mystery was finally solved by physicists at the turn of the twentieth century.

Light as a particle

In the latter years of the nineteenth century, a lot of effort was being devoted to the study of thermal radiation. This is electromagnetic radiation which has been emitted by a hot object. It was found that, although the hot object emitted radiation across the entire electromagnetic spectrum, the intensity of that radiation varied with wavelength. The wavelength at which most of the energy was emitted also varied according to the temperature of the object. In deriving a mathematical expression which would reproduce the curves of

Figure 1.8
Contimuum, absorption and emission spectra

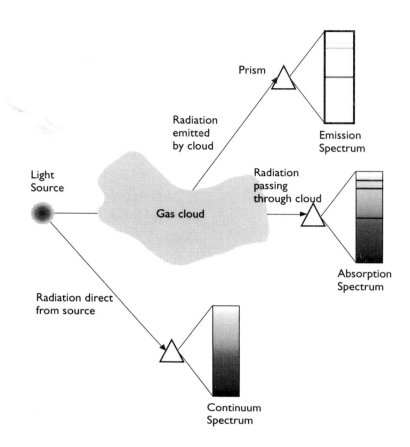

intensity against wavelength, German physicist Max Planck had to make a radical assumption. This was that the radiation, visible light included, released by the atoms of the hot object could only carry certain pre-defined energies. This censorship of energies flew in the face of contemporary science but was the only way in which Planck could make his formula work. The concept of discrete energies became known as quantization and laid the foundation stone of today's quantum theory.

By 1905 the concept of quantization was so widely used that Einstein proposed a remarkable idea. Even though it was accepted that light was a wave motion, he assumed that, under some circumstances, light could behave like a stream of particles. Each of these particles would carry an amount of energy which

Mathematical note 1.5

Wavelength and energy of a photon

If light has a frequency, f, the energy of each photon is given by:

$$E = hf$$

where
E = energy of a photon
h = Planck's constant
f = frequency

Since,

$$c = f\lambda \qquad E = \frac{hc}{\lambda}$$

where
c = speed of light
λ = wavelength

Figure 1.9
Wave-particle duality

This diagram can be seen as either a vase or two faces looking at each other. Never can it be seen as a vase and two faces. When you visualise the faces, the vase disappears. In a similar way, light can be imagined as either a wave or a particle but never both at once.

was inversely proportional to its wavelength. In other words, the longer the wavelength of radiation, the less energy these particles carry (see *mathematical note 1.5*). Einstein used this idea to explain the photoelectric effect and the concept of light behaving as particles became established. The particles of light were later called photons but it was still impossible to explain the propagation of light if it were particles, so the wave models persisted as well. To reconcile the two, seemingly contradictory ideas about light, the concept of wave-particle duality was born. This states that under certain conditions light behaves as if it were a wave but in other circumstances it appears as if it were a particle. The mechanics of light's propagation and the way in which light rays interact with one another is understandable only by use of wave models. The way in which light interacts with matter becomes understandable only if light is thought of as a particle. It is impossible to think of light being both a wave and a particle at the same time (see *figure 1.9*). Instead, it must be recognised when to use a wave model and when to use a particle model.

In order to understand how spectral lines are formed it must be assumed that light is a particle: a photon. It was proposed by Ernest Rutherford in 1911 that atoms contain a massive, positively charged nucleus. The majority of an atom's volume is therefore empty space in which the negatively charged electrons exist. The arrangement of those electrons was unknown until Niel Bohr used the concepts of quantization to explain that the formation of hydrogen's spectral lines was directly related to how electrons are organised around hydrogen nuclei. Bohr's superlative work on the structure of the hydrogen atom led the way to a complete understanding of the mechanisms by which atoms absorb and emit discrete wavelengths of electromagnetic radiation.

The key to understanding why atoms produce spectra with characteristic patterns of lines is that the spectral lines

Figure 1.10
Energy levels around a hydrogen atom

Electron transitions to and from specific energy levels are called series. The Balmer series of lines occurs in the visible region of the electromagnetic spectrum. Each level contains electrons of a specific energy. The higher the number, the more energy needed by the electron to reach that state.

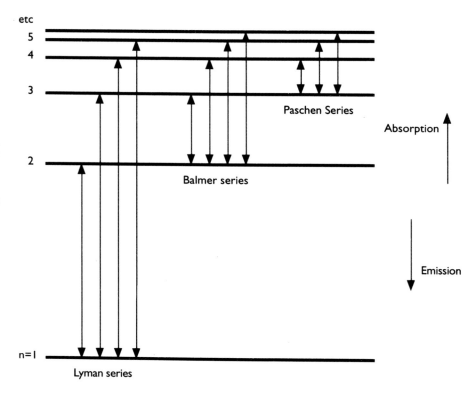

correspond to the arrangement of electrons around the atom's nucleus. The electrons can only exist in certain energy levels. The type of chemical element to which the atom belongs defines the energy levels. Since each level corresponds to a specific energy, in order for an electron to move from one level to another, energy must be either lost or gained. That difference in energy is made up by the absorption or the emission of a photon which contains that energy. Only photons which contain the precise amount of energy needed to make a transition are absorbed or emitted. Hence, spectral lines appear at set positions within a spectrum.

In a hydrogen atom, certain transitions affect photons of light in the visible region of the electromagnetic spectrum. These transitions are known as the Balmer series (see *figure 1.10*) and correspond to transitions involving the second energy level and above. For example, an electron jumping from the third energy level to the second will emit a photon of light which, because of the amount of energy it carries, appears as a ray of red light with a wavelength of 656 nanometres.

Bohr used his theory of electron energy levels to predict the existence of a series of spectral lines in the ultraviolet region of the spectrum. These were based upon transitions from the first energy level and above. The Lyman series, as they became known, were duly discovered at exactly the correct wavelengths Bohr's model predicted.

By thinking of the arrangement of electrons around atomic nuclei, the reason why atoms only absorb or emit certain wavelengths of light becomes obvious. It is because only those photons which contain enough energy to loft the electrons to another energy level are absorbed. When the electrons return to their original levels they emit that energy as photons. Hence, in an absorption spectrum of hydrogen, the dark line at 656 nanometres corresponds to photons which have been absorbed by electrons making the transition from the second energy level to the third. The 486 nanometre line, in the blue region of the spectrum, corresponds to transitions between the second energy level to the fourth. The more energy required to make the transition, the shorter the wavelength of radiation involved.

Emission takes place when an electron jumps from a higher energy level to a lower energy level. Thus, the spectral lines in atomic emission spectra occur at the same wavelength as the spectral lines in atomic absorption spectra.

Redshift

It is the study of spectral lines which led to the discovery of the redshift phenomena (see figure 1.11). Under certain external conditions, which this book will explore, the wavelength of spectral lines can be altered. This alteration is produced in one of three ways. The first way is by the motion of either the

Figure 1.11
The phenomenon of redshift

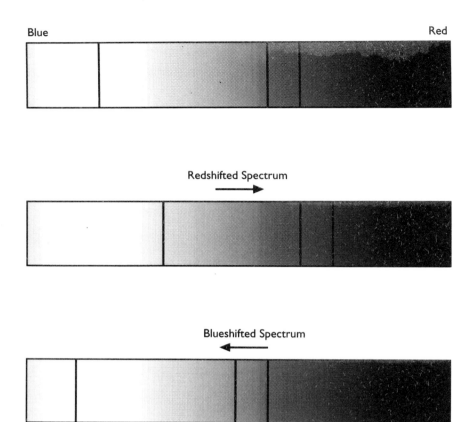

Blue

Red

Redshifted Spectrum

Blueshifted Spectrum

source of radiation or the object which observes it. The second way is when radiation is emitted by an object with an intense gravitational field and the final way is when radiation is stretched by the expansion of the universe. The phenomenon gets its name because, in most cases, the spectral line has its energy decreased which causes its wavelength to increase. Any increase in the wavelength of visible light will make it appear redder than it was originally. It will appear to move towards the less energetic end of the electromagnetic spectrum. Conversely a mechanism which squashes the light will make it appear bluer. This rarer phenomenon is termed blueshift.

Observing redshift is actually quite simple. It requires that a spectrum of the celestial object in question be taken, followed by a means of measuring any shift in the position of its spectral lines. For something whose measurement is this simple, the implications of redshift, as we shall see, are many, varied and sometimes profound. For example, the cosmological redshift phenomena caused by the expansion of the universe underpins one of the most important theories of the twentieth century: the big bang theory of how the universe was created. If this redshift were to be explained in another way, as will be explored in Chapter Five, our entire view of the cosmos' origins would be destroyed!

2

The Doppler effect

Christiaan Doppler was an Austrian physicist whose distinguished career took him to the directorate of the Vienna Physical Institute. On the way he also became a professor at the University of Vienna but Doppler is best remembered for his work concerning sound waves emitted and received by moving objects. He described the principle whereby the pitch of a sound is altered depending upon the way in which the emitting object is moving.

Two pieces of terminology which will be used repeatedly in this chapter are source and observer. A source is any object which is emitting a wave motion, eg a torch emitting light or a loud speaker emitting sound. An observer is any object or person, in receipt of a wave motion.

Imagine a transmitter regularly sending out pulses. If the transmitter were stationary, the length of time between pulses would be measured and found to be the same, no matter where the observer stood in relation to the transmitter. Imagine next what would happen if the transmitter began to move. In the example from Chapter One it was seen that the time between pulses increases for an observer if the source is moving away. It is therefore no surprise that the pulse time appears to shorten if the source is travelling directly towards the observer (see figure 2.1).

If, in place of discrete pulses, we next imagine that the transmitter is giving out a continuous signal, then what were once pulses are now crests in a wave motion. The timing between crests gives the frequency of the wave and, providing the speed of its propagation is known, the wavelength can be easily calculated as well.

Figure 2.1 represents a wave released by two sources. In the top image, the source is stationary but in the bottom image, the source is moving. The circles represent the wave crests and so the spacing between them shows the wavelength. In the image of the moving source, wave crest 'A' was released when the source was at position 'a', wave crest 'B' was released when the source was at position 'b' and so on. In this image it is obvious that observers in

Figure 2.1
The Doppler effect

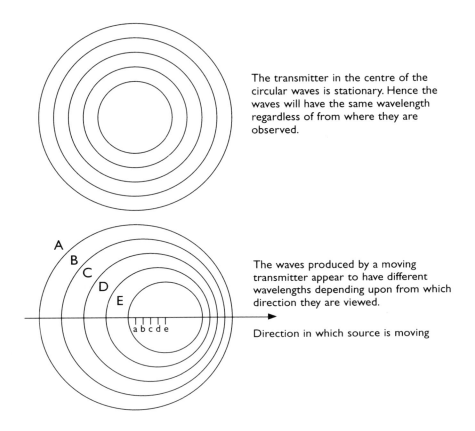

The transmitter in the centre of the circular waves is stationary. Hence the waves will have the same wavelength regardless of from where they are observed.

The waves produced by a moving transmitter appear to have different wavelengths depending upon from which direction they are viewed.

Direction in which source is moving

the direction of the source's motion will observe emission which has had its wavelength compressed (blueshifted) whilst observers behind the source will notice that the wavelength has been stretched (redshifted).

If the source is emitting sound waves, the pitch is affected. It will be raised for blueshifted emission and lowered for redshifted emission. If the source is light, the energy of its photons is changed. The photon energy will be increased for blueshifted emission and decreased for redshifted emission.

Although we have talked about moving sources and stationary observers, since all motion is relative, the precise situation could be reproduced if the source were stationary and the observer were moving. Equally likely is that both source and observer are moving. In this case, the velocities of both are added together and it is the resultant velocity which determines the extent of the Doppler shift.

The precise way of calculating the wavelength shift caused by a moving object is to multiply the wavelength of emitted radiation by the ratio of the object's velocity to the speed of light. This ratio is given the letter 'z' and is constantly referred to in astronomy. The 'z' ratio is also calculable by taking the ratio of the wavelength shift to original wavelength. (see *mathematical note 2.1*).

If the radiation from an astronomical source is split into a spectrum, the way in which the brightness varies along the spectrum is known as the energy distribution. A redshift does not move the spectrum 'en masse', instead it stretches this energy distribution. The wavelength shift produced by a velocity is smaller for high energy photons (i.e. those with short wavelengths) than it is for low energy photons (i.e. those with long wavelengths). So, although all photons in the spectrum are reduced in energy, the high energy photons lose less energy than the low energy photons. In fact, if the wavelength of photons in question is doubled, so are their respective wavelength shifts (see *figure 2.2*).

This aspect of the phenomenon is obvious if it is assumed that the velocity is entirely due to the motion of the source away from the observer. Remembering that this is how it would appear from the observer's frame of reference, imagine two wavelengths of radiation being emitted by the same moving body. If one of the wavelengths is double that of the other, it takes twice the length of time for it to be emitted. Since it is the motion of the source away from the observer which stretches the radiation, the shift in the wavelength is exactly equal to the distance that the object has moved during the time taken to emit each wavelength. Hence, if one wavelength is twice as long

Mathematical note 2.1

The Doppler effect

To obtain a mathematical formula for the wavelength shift induced by radial motion, consider the example of the two cars measuring light pulses from figure 1.7. It was shown that:

$$T_A = (t' - t) + \left[\frac{(d' - d)}{c} \right]$$

$$T_B = (t' - t)$$

where
T_A = A's measured pulse time
T_B = B's measured pulse time
t = time of first pulse
t' = time of second pulse
d = distance between A and B when the first pulse is released
d' = distance between A and B when the second pulse is released
c = speed of light

Combining these two equations gives:

$$T_A = T_B + \left[\frac{d' - d}{c} \right]$$

If the pulses are successive wave crests in a ray of light, then T_A and T_B can be used to give the wavelength of light in the following way:

$$\lambda = cT_A$$

$$\lambda_o = cT_B$$

where λ = observed wavelength
λ_o = emitted wavelength

Next, the distances d' and d must be replaced, using the simple formula:

$$d = v_r t$$

$$d' = v_r t'$$

where
v_r = radial velocity of the source

Hence our equation becomes:

$$\lambda c = \lambda_o c + \frac{v_r (t' - t)}{c}$$

which simplifies to:

$$\frac{\lambda}{\lambda_o} = 1 + \frac{v_r}{c}$$

This equation is often presented in the form of a ratio between the wavelength shift, $\Delta\lambda$, and the emitted wavelength λ_o. This means it has to be slightly recast because $\lambda = \lambda_o + \Delta\lambda$.

It becomes:

$$\frac{\Delta\lambda}{\lambda_o} = \frac{v_r}{c}$$

These ratios are often referred to as the parameter z.

as the other, it will take twice as long to be emitted, the source will travel twice as far and the shift in its wavelength will be

Figure 2.2
Redshifted energy distribution

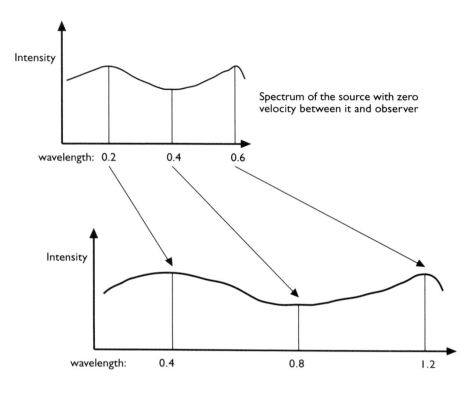

Intensity

Spectrum of the source with zero velocity between it and observer

wavelength: 0.2   0.4   0.6

Intensity

wavelength: 0.4   0.8   1.2

Spectrum of the source with a velocity corresponding to z=1 between it and observer

twice as large. Its redshift, however, as denoted by 'z' will remain the same because the velocity of the source has not changed.

Redshift is a subtle phenomena when applied to a continuous spectrum. This type of spectrum is emitted by a hot object, such as a solid, liquid or a dense gas and displays no distinguishing features. The energy distribution of a spectrum is indicative of the temperatures of the emitting object. Thus, any redshift imparted to a continuous spectrum simply alters the apparent temperature of the source from which it was emitted. Luckily, almost all objects in the universe emit radiation which contains absorption and/or emission lines in their spectra. Since each chemical element has its own characteristic pattern of these lines, it makes it much easier to spot if the lines have been stretched out towards the red end of the spectrum or squashed up towards the blue.

To assist in making this determination, and to allow measurement of the wavelength shifts, astronomical spectrometers are fitted with the ability to place a few emission lines above and below the observed spectrum. These reference lines are of known wavelengths and so allow the displacement of the lines in the observed spectrum to be measured and their new wavelengths calculated.

Radial velocity

The velocity calculated by measuring the redshift or blueshift of an object is known as its radial velocity. This is because it is caused by motion along the line of sight radially away from, or towards, the observer. Almost certainly, however, the source is not just moving radially with respect to the observer; nature is seldom that easy. Instead, the radial velocity is usually a component of the actual motion of the source.

The transverse direction is at right angles to the radial direction and corresponds to movement across the night sky. If the velocity in the transverse direction is known, trigonometry can be used to combine the radial and transverse components into the actual velocity that the object has with respect to the Earth.

One of the easiest ways of calculating the transverse velocity of a nearby celestial object, such as a star, is to measure the difference in its position on the night sky after a set period of time. This quantity is known as the proper motion and is measured in arc seconds per year. It can be used in conjunction with the parallax of the star in question to give a figure for the transverse velocity (see mathematical note 2.2).

The convention in astronomy is to refer all stellar motions to what is called the local standard of rest. This is a frame of

Mathematical note 2.2

Transverse and radial components of velocity

Calculating the velocity of a star with respect to the Earth, the radial velocity is given by

$$v_r = \frac{c\Delta\lambda}{\lambda_o} = cz$$

where
v_r = radial velocity
c = speed of light
$\Delta\lambda$ = wavelength shift
λ_o = original wavelength
z = redshift

The transverse velocity is given by

$$v_t = 4.74\frac{\mu}{\pi}$$

where
v_t = transverse velocity (m/s)
μ = proper motion (arc seconds per year)
π = parallax (arc seconds per year)

Since the radial and transverse velocities are at right angles to one another, they represent components of the actual motion

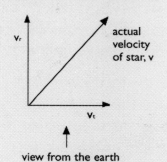

It is obvious that the actual velocity of the star can be given by Pythagoras' theory such that:

$$v = \sqrt{\left(v_r^2 + v_t^2\right)}$$

reference which is centred upon the Sun. The local standard of rest is in orbit around the centre of the Galaxy and has a velocity which has been calculated by averaging the motion of all the stars in the solar neighbourhood. This gives the local standard of rest a velocity of 250 km/s with respect to the centre of our Galaxy. Even though it is centred on the Sun, because its velocity has been calculated as an average of those of the nearby stars, its velocity is different from that of the Sun by some 20km/s. The local standard of rest is a more physically satisfying frame of reference to use than the Sun's frame of reference. Any deviation in the velocities of stars from the local standard of rest are referred to as peculiar velocities.

Rotation of the Galaxy

By studying the way in which stars move in relation to the local standard of rest, it is possible for astronomers to deduce the manner in which the galaxy rotates. It has been observed that the Doppler shifts of stars around the solar neighbourhood are distributed into a quadrupole. Stars which are 'behind' the Sun and closer to the centre of the Galaxy exhibit redshifts, as do stars which are further away from the Galactic centre but 'in front' of the Sun. Blueshifted stars are those which are behind the Sun but further away from the Galactic Centre and those in front of the Sun but closer to the centre of our galaxy. These observations can be converted into peculiar motions and placed in a schematic diagram (see figure 2.3). From this diagram it becomes immediately obvious that stars closer to the centre of our galaxy have orbital velocities which are slower than the Sun whilst those further out are travelling faster in their orbits.

If this method could be extended beyond the solar neighbourhood, it would be possible to map the entire galaxy and see exactly how it rotated. Unfortunately, this is impractical because the Galaxy contains a lot of obscuring material, such as

Figure 2.3
Stellar movement in the solar neighbourhood

Although the actual motion of stars in the solar neighbourhood is that stars closer to the galactic centre move faster than those further away, observations from Earth have to take into account the Sun's motion. Effectively, the Sun appears stationary to us which causes its motion to be subtracted from the stars around us. This means that the apparent motion of the stars in the solar neighbourhood gives a pattern of redshifts and blueshifts which alternate between the quadrants surrounding the Sun.

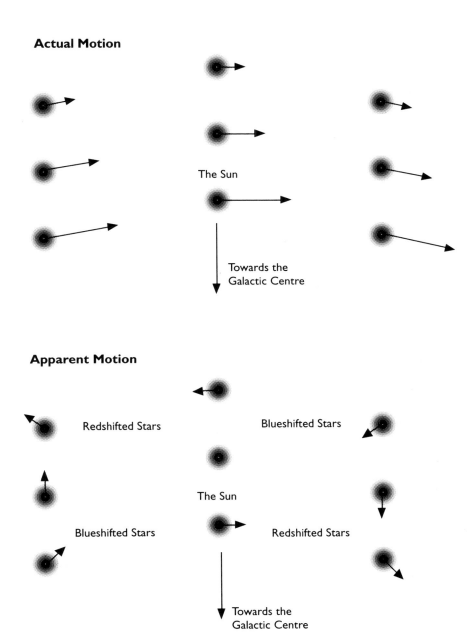

Actual Motion

The Sun

Towards the Galactic Centre

Apparent Motion

Redshifted Stars

Blueshifted Stars

The Sun

Blueshifted Stars

Redshifted Stars

Towards the Galactic Centre

dust clouds, which block the light from distant stars and make them impossible to observe.

In 1931, however, Karl Jansky was conducting some research into 'static' at the Bell Telephone Laboratories using radio receiving equipment. He discovered that the Galaxy radiates at radio wavelengths. Although his work was largely ignored, astronomers found vastly superior radio technologies available to them, thanks to the experiments with radar after the cessation of World War II hostilities in 1946. Upon re-examining the Jansky emissions, the science of radio astronomy was born.

One of the people who had picked up on the potential importance of Jansky's work was the Dutch astronomer H.C. van de Hulst. In 1944, Hulst predicted that it would be possible to detect the hydrogen in the galaxy with the use of appropriate radio receivers. Although most of the atomic transitions in hydrogen cause emission in the ultraviolet and optical, which is blocked by the interstellar clouds, Hulst correctly pointed out a transition which would emit in the radio region of the spectrum. It was called the spin-flip transition.

Quantum theory ascribes certain properties to the particles which make up the universe. These properties are described by a set of quantum numbers, which are part of each particle's vital statistics. They tell us in which quantum state the particle exists. When electrons cause emission or absorption, by jumping between energy levels, all they are doing is using the energy in absorbed or emitted photons to change their quantum states. The situation is made slightly more complicated by the fact that, around an atomic nucleus, no two electrons can exist in the same quantum state. It is actually known to physicists as Pauli's Exclusion Principle.

One of the properties described by the quantum numbers is that of spin. Electrons can exist in the ground state of hydrogen with spin up or spin down. If the electron spin is compared to that of the proton, which forms the hydrogen nucleus, it can be either parallel, if both are spinning in the same direction, or anti-parallel, if they are spinning in different directions. A parallel spinning electron contains slightly more energy than an anti-parallel spinning one.

The laws of quantum mechanics allow particles to behave spontaneously. It is therefore entirely possible for the electron to change spontaneously from a parallel spinning state to an anti-parallel spinning state. Since the anti-parallel state is slightly less energetic than the parallel state, the excess energy is emitted as radiation with a wavelength of twenty one centimetres. Although the probability of this occurring in each atom is only once in every eleven million years, the galaxy contains countless billions of hydrogen atoms and so the emission is constant. The wavelength of the radiation is so long that it can pass unhindered through the clouds of material which block the visible light. By detecting the hydrogen emission it should be possible, in principle, to map the entire galaxy.

The down side of the fact that the emission can pass through the obscuring clouds of material which exist within our galaxy is that any detection is a confusion of signals from hydrogen clouds at different distances along the line of sight of the radio telescope. Thankfully, the Doppler effect causes different wavelength shifts to be imparted upon the twenty one centimetre emission from each individual cloud. The individual wavelength shifts depend upon the clouds' locations relative to the centre of the galaxy (see figure 2.4). This allows the clouds' radial velocities to be calculated and the distribution of matter in the Galaxy to be elucidated.

Analysis of cloud motions also helps astronomers to understand the way in which the Galaxy rotates. By taking the various observations into account, it can be seen that the orbital velocity of matter in our galaxy increases with distance from the centre (see figure 2.5). The expected result was that the velocity would drop off with distance like the velocity of the planets in our solar system. This is by no means the case for the galactic rotation curve and so an immediate conclusion of the Doppler mapping is that there is a substantial amount of celestial material which is still awaiting discovery in the outer regions of our galaxy. The

Figure 2.4
Hydrogen cloud spectrum

The hydrogen emission from clouds in the galaxy has been redshifted by varying amounts depending upon the cloud's distance from the galactic centre. λ_0 known as the rest wavelength, marks the position at which the hydrogen emission line would have occured if the clouds had not been in motion. The height of the emission lines signifies the strength of radiation from each cloud.

(NASA and STScI)

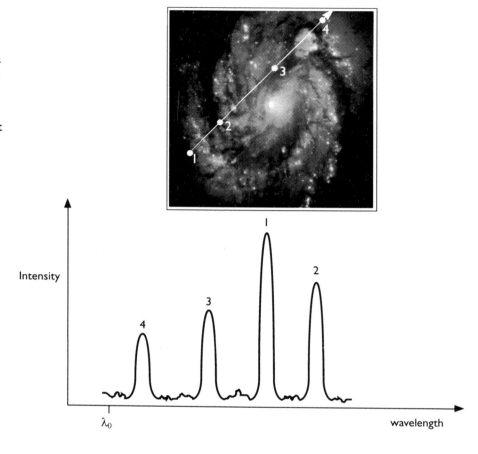

Figure 2.5
Galactic rotation curve

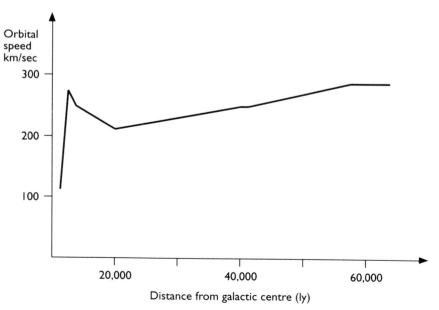

Distance from galactic centre (ly)

precise nature of this material has been the subject of much debate and has been popularised by being given the name dark matter. It has achieved a nearly mystical status in some ways and will be briefly discussed in Chapter Five.

Spectroscopic binaries

As well as studying the orbital motion of stars around the galactic centre, the Doppler effect can be used to study the orbital motion of stars around each other. Approximately fifty per cent of the stars in the galaxy are binary or multiple stars. This means that half the galaxy's stars are not just single entities like the Sun. Instead, they are gravitationally bound to one, or more, other stars. In many cases, just by looking through a telescope it is obvious that what appear to be single stars are, in fact, double stars. Sometimes, however, the stars are so close together that, even through the most powerful telescopes, they still appear to be single stars. In cases like this, spectroscopy can often differentiate between the two stars.

Taking the spectrum of a star is a useful way to classify its properties. Sometimes these stellar spectra display two set of spectral lines. One set has been redshifted from its rest wavelength, the other set has been blueshifted. If the star is observed over a period of time, it will be noticed that the spectral lines oscillate about an average position. If this is observed, it means that the star is actually a binary star. As the two stars orbit one another, so their radial velocities change, causing the Doppler shifts to alter. This phenomenon manifests itself as the oscillating spectral lines (see figure 2.6).

Doppler broadening

All emission and absorption of spectral lines takes place via the interaction of radiation with the atoms in gas clouds. So far we have treated spectral lines as being very thin, uniform things

Figure 2.6
Spectroscopic binary stars

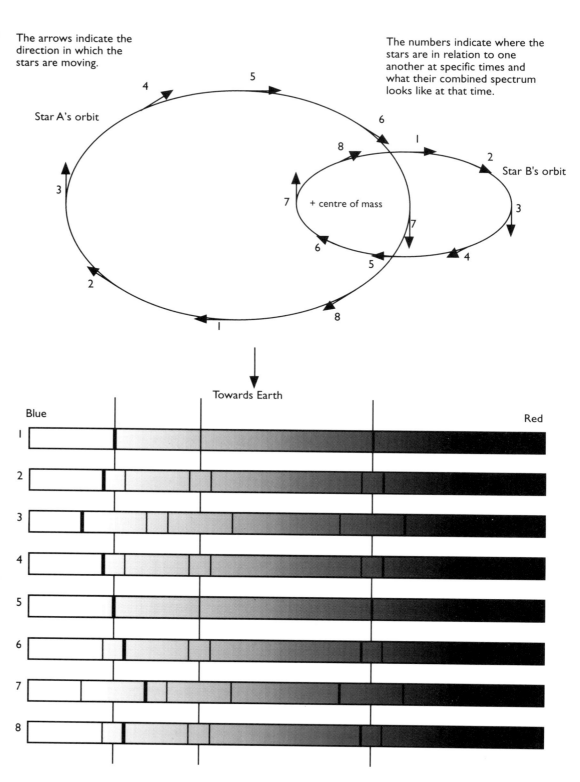

The arrows indicate the direction in which the stars are moving.

The numbers indicate where the stars are in relation to one another at specific times and what their combined spectrum looks like at that time.

Star A's orbit

Star B's orbit

+ centre of mass

Towards Earth

Blue

Red

Figure 2.7
Spectral line profiles

Idealised Line Profile

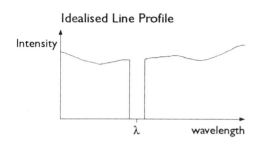

Gaussian Line Profile

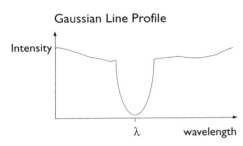

which occur at a single wavelength. However, the atoms of gas in the cloud all exhibit some motion and this thermal motion of atoms occurs in random directions, causing some to move towards the observer and others to move away. The precise range of velocities that the atoms possess varies according to the temperature of the material. This motion causes a Doppler effect to be imparted to the emission or absorption from each and every atom in the cloud. This distorts the profile of the line from a narrow rectangle into a curve known as a Gaussian profile *(see figure 2.7)*. The precise width of the Gaussian depends upon the temperature of the gas cloud *(see mathematical note 2.3)*.

Any process which disturbs the atoms into motion results in spectral lines being broadened because of the Doppler effect. Doppler broadening is the generic term for all these mechanisms. For instance, in addition to the individual thermal motions of the atoms, spectral lines can also be Doppler broadened because of turbulence within the cloud of gas *(see mathematical note 2.4)*.

Rotation of the object is often the predominant broadening mechanism in the observation of stars. The stars are so far away that even though some are enormously big, they appear as point sources to us. As they rotate, the light from one limb is blueshifted as that limb approaches, and the light from the other limb is redshifted as it recedes. Since the star is an unresolved point source, however, (i.e. it is impossible to distinguish one limb from the other) the red and blueshifted emission is merged into a single large spectral line with a hemispherical profile *(see mathematical note 2.5)*.

Rotation can also take place in objects which are resolvable. For instance, a planet reflecting sunlight can be seen as a disc rather than just a point source. The act of reflection from a moving body superimposes a Doppler shift just as certainly as if the

Mathematical note 2.3

Thermal motions and the profile of spectral lines

The Gaussian profile is described by the equation

$$I(f) = \frac{I(f_o)}{f_o} \sqrt{\left(\frac{mc^2}{2\pi kT}\right) \left[\frac{-mc^2}{2kT}\left(\frac{f-f_o}{f}\right)^2\right]}$$

where
$I(f)$ = the intensity at frequency,
$I(f_o)$ = the intensity at the rest frequency, fo
m = the mass of an atom in the cloud
c = the speed of light
k = the Boltzmann constant
T = the temperature of the emitting cloud

When measuring the width of a spectral line, the convention is to measure the width at half the maximum intensity. This solves the messy experimental problem of measuring to the very edges of the Gaussian profile. Often the outer extremities are lost in the surrounding continuum radiation, making the full width of the line impossible to measure. The width of the line at half maximum is usually abbreviated to FWHM (full width, half maximum), and is given mathematically by the equation:

$$FWHM = \frac{2}{f_o}\sqrt{\left[\frac{2kT}{m}\ln 2\right]}$$

Mathematical note 2.4

Turbulence and the profile of spectral lines

In the case of turbulent motion within the cloud, the Gaussian Profile is described by the equation:

$$I(f) = I(f_o)\frac{1}{\sqrt{\pi}}\frac{c}{Vf}\exp\left[-\frac{c^2}{V^2}\left(\frac{f-f_o}{f}\right)^2\right]$$

where
$I(f)$ = the intensity at frequency, f
$I(f_o)$ = the intensity at the rest frequency, f_o
c = the speed of light
V = the average of all the velocities contributing to the turbulence

In this circumstance, the FWHM is given by:

$$FWHM = \frac{2}{f_o}\sqrt{\left[\left(\frac{2kT}{m} + V^2\right)\ln 2\right]}$$

In most cases, thermal and turbulent Doppler broadening are present and so the two equations can be combined to give a single equation which describes the shape of the Gaussian:

$$I(f) = I(f_o)\frac{c}{\sqrt{\left[\left(2\pi k\frac{T}{m}\right) + V^2\right]}}f^{-1}\exp\left[\frac{-c^2}{\sqrt{\left[\left(2k\frac{T}{m}\right) + V^2\right]}}\left(\frac{f-f_o}{f}\right)^2\right]$$

Mathematical note 2.5

Rotation and the profile of spectral lines

The spectral line profile for an unresolved, spherical source which is rotating is given by:

$$I(f) = I(f_o)\sqrt{1 - \frac{c^2(f - f_o)^2}{V^2 f^2}}$$

where
$I(f)$ = the intensity at frequency, f
$I(f_o)$ = the intensity at the rest frequency, f_o
c = the speed of light
V = the average of all the velocities contributing to the turbulence

The profile is that of a hemisphere:

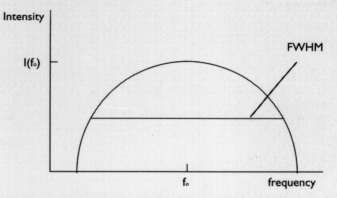

The width of the line profile at full width, half maximum intensity if given by:

$$FWHM = \left(\sqrt{\frac{3}{2}}\right)\left(\frac{V}{f_o}\right)$$

moving object had emitted the radiation itself. If a spectrum were taken across the equator of a rotating planet, the resultant spectral lines would be diagonal because from one limb of the planet they are being blueshifted and from the other limb they are being redshifted (see figure 2.8).

Certain stars, such as some classes of variable stars, expand and contract. The expansion of the stars' outer layers blueshifts the radiation being emitted. This blueshift is different across the face

Figure 2.8
Diagonal spectral lines

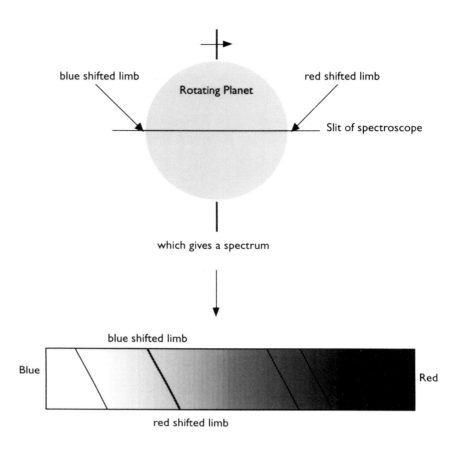

Figure 2.9
Stellar expansion

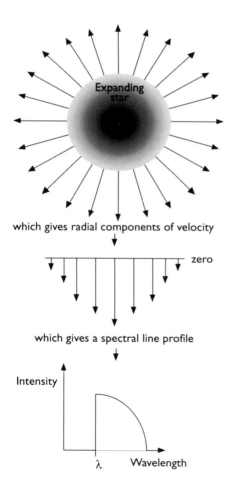

of the star, however, because only at its centre is the expansion completely in the radial direction *(see figure 2.9)*. Redshifts are caused when the surface of the star contracts.

Stars can sometimes throw off shells of gas which drift into space. The slow moving gas in these expulsions stays close to the stellar surface and is heated by the star's radiation. These clouds then emit spectral lines, which are only slightly blueshifted. The faster moving gas has travelled further from the star and has become cooler because it is further away. The radiation emitted by the inner gas must travel through this material before it escapes to space. As it does so, some of it is absorbed. Since this gas is travelling faster than the emitting material, the absorption line does not correspond to the same

Figure 2.10
P Cygni spectral line profiles

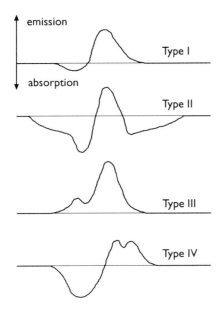

wavelength. Instead, it is blueshifted more dramatically, which causes the emission line and the absorption line to appear adjacent to one another on the spectrum *(see figure 2.10)*. These are known as P Cygni line profiles and derive from one of the variable stars in the constellation Cygnus, whose spectrum displays these profiles.

Active galaxies can be classified according to the extent of the Doppler broadening which their spectral lines have suffered. About one in ten galaxies is 'active', which means that some kind of extraordinary activity is taking place in its centre. The most widely accepted theory states that a massive black hole lies at the centre of each and every active galaxy, around which clouds of hydrogen gas swirl turbulently. They are heated so much by the material falling into the black hole that they emit radiation in the form of spectral lines. Surrounding the equatorial region of the active galactic nucleus is a large dusty torus which blocks the view of the fastest moving hydrogen clouds from certain viewing angles. The region containing the hydrogen clouds is nominally split into two sections, the so-called broad line region (BLR) and the narrow line region (NLR). The broad line region corresponds to deep down in the nucleus, very close to the black hole, where the hydrogen clouds are moving the fastest. Here the clouds are orbiting at velocities of 10,000 km/s, which Doppler broadens the emission lines considerably. If our view of this region is blocked by the dust torus, it may be possible to see only the narrow line region. The narrow emission lines are produced by hydrogen clouds further away from the black hole which orbit at velocities of about 1,000 km/s.

A popular theory in the classification of these active galaxies is that the various different classes may all be the same type of object but simply viewed from different angles. A slightly more modest version of this idea is that all narrow line active galaxies have a hidden broad line region. These hidden broad lines are made visible by filtering out everything except the scattered radiation.   Although this theory is by no means proven, one of the best examples of this idea is Seyfert galaxy NGC 1068. Seyfert galaxies come in two varieties: type 1 with broad lines and type 2 with narrow lines.   Originally, this galaxy was classified as a Seyfert type 2, a narrow line galaxy, but observations in scattered light show that it has a hidden broad line region, meaning that it is an obscured Seyfert type 1. *(See figure 2.11).*

Relativistic Doppler effect

One of the predictions of Einstein's special theory of relativity was that movement twists one's perceptions of the universe. Chapter One illustrated this by using an example of two cars, one in motion and one stationary, measuring the time between flashes of a torch which was carried on the moving car. It was shown that the time measured by the stationary car was longer than that measured by the moving one. So, to outside observers (i.e. those in other frames of reference), it has been shown that time appears to pass slowly in moving frames of reference than it does in stationary ones. After further investigation, it would be found that the faster the frame of reference is moving, the more pronounced is the effect of time slowing down. It is a phenomenon known as time dilation and is one of the most celebrated of special relativity's predictions. At velocities of less than ten per cent of the speed of light, the effect is negligible. At velocities greater than ten per cent of the speed of light, however, special care must be taken to allow for time dilation. *(See figure 2.12).*

The Doppler shift imparted on radiation is dependent upon the observed time between the emission of successive wave crests. If the difference in velocity between the source and observer is relativistic, i.e. a velocity at which time dilation will become significant, the wavelength shift imparted on the radiation will be increased from the straight forward, non-relativistic ratio of velocity to the speed of light. This is because outside observers

Figure 2.11
An obscured Seyfert type 1 nucleus

The gas in the accretion disc is heated so much that it emits x-rays. These excite the hydrogen clouds in the BLR and produce broad spectral emission lines. The broad spectral lines are not able to escape through the torus into Earth's line of sight but are scattered from the narrow line producing clouds. This means that if the much more powerful narrow line radiation is screened out, the broad lines suddenly become visible.

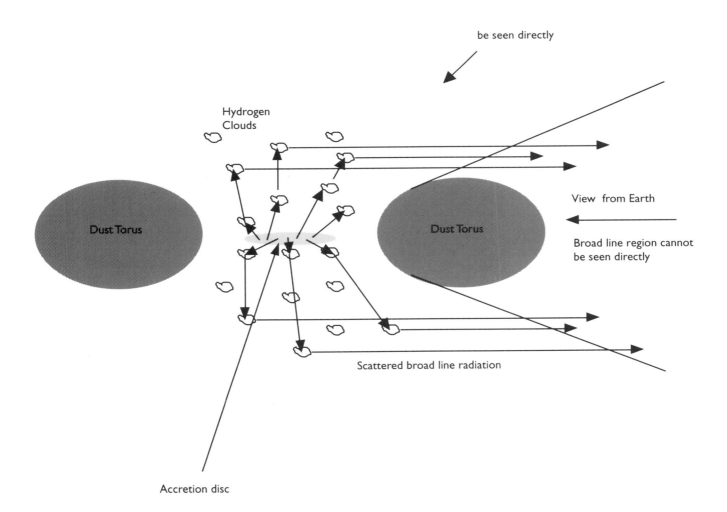

Figure 2.12
Classical and relativistic Doppler shift

This diagram shows that between zero and ten percent of the speed of light, the classical form of Doppler effect is sufficient. At velocities greater than ten percent of the speed of light, the relativistic form of the Doppler shift equation must be used.

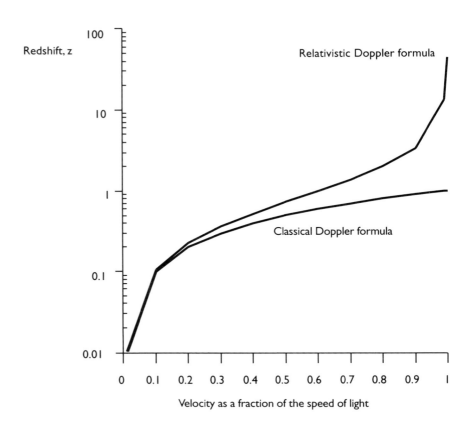

are given the impression that time has slowed down for the object in motion. If the moving object releases a burst of radiation, observers will see it being emitted over a longer period of time than is usual. Since the emission takes place for longer, it means the frequency of the radiation will decrease and hence its wavelength will increase – causing a redshift.

The mathematical method of mapping the observations from one frame of reference into another is known as the Lorentz transformation. This is for historical reasons: it was Lorentz who proposed that, perhaps, length is contracted in the direction of motion. He did this in order to explain away the failure of the Michelson-Morley experiment to detect the ether. Despite lacking a real physical basis in which to couch his assumptions, Lorentz gamely developed a set of transformation equations which would alter the length by exactly the right amount in order to 'fiddle' the results of the experiment. Later, when Einstein was working on his special theory of relativity, he used his two postulates of relativity to determine how objects in frames of references are seen by outside observers. The interesting result that he found was that to an outside observer an object in a moving frame of reference would indeed display differences in the length, mass and passage of time. Upon advancing his work to calculate the magnitude of these differences he found they were exactly equal to those derived by Lorentz *(see mathematical note 2.6)*. Lorentz had derived empirically what Einstein discovered theoretically. The Lorentz transformation can be used in order to calculate the exact magnitude of the time dilation effect *(see mathematical note 2.7)*.

So, at relativistic velocities the Doppler effect has two components: the first is the conventional wavelength shift caused by the movement of the source or observer, as introduced in Chapter One and elaborated upon at the beginning of this chapter. This wavelength shift can be either a blueshift or a redshift depending upon whether the distance between the source and observer is increasing or decreasing.

Adding to this shift, however, is a second component, caused by time dilation. The wavelength shift caused by time dilation is always a redshift – even if the source is travelling towards the observer. If the motion of the object is purely in the line of sight, then a blueshift still occurs because the motion of the object towards us overwhelms the time dilation effect *(see figure 2.13)*. In some cases, however, objects which are travelling with both radial and transverse component to their motion can display redshifts even if their radial component of motion is towards us *(see figure 2.14)*.

Mathematical note 2.6

The Lorentz transformation

Imagine two frames of reference, S and S', which are moving with respect to each other at a velocity of v. An event takes place which has co-ordinates (x,y,z,t) in S and (x',y',z',t') in S':

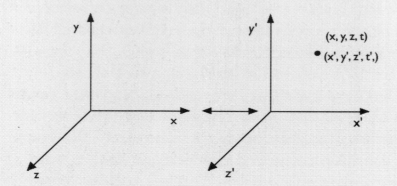

From the diagram it can be easily seen that: $y = y'$
$$z = z'$$

Placing the frames of reference onto a spacetime diagram gives the following:

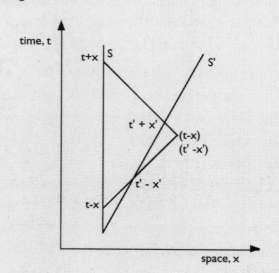

By referring back to mathematical note 1.4 where the k factor was derived it should be seen that:

$$t' - x' = k\left(t - x\right)$$
$$t + x = k\left(t' + x'\right)$$

Remembering that in relativistic units:

$$k = \sqrt{\frac{1 + v}{1 - v}}$$

Substituting and rearranging gives the Lorentz transformation in relativistic units as:

$$t' = \frac{t - vx}{\sqrt{1 - v^2}} \quad x' = \frac{x - vt}{\sqrt{1 - v^2}} \quad \begin{array}{l} y' = y \\ z' = z \end{array}$$

In non-relativistic units:

$$k = \sqrt{\frac{c + v}{c - v}}$$

So the Lorentz transformation equations become:

$$t' = \beta\left(1 - \frac{vx}{c^2}\right) \quad x' = \beta\left(x - vt\right) \quad \begin{array}{l} y' = y \\ z' = z \end{array}$$

where   c = the speed of light

$$\beta = \frac{1}{\sqrt{\left(1 - \frac{v^2}{c^2}\right)}}$$

Note that when v →0, β →1 and relativistic effects become negligible. In practice this happens when v is less than ten per cent of the speed of light.

Mathematical note 2.7

Time dilation

Imagine two frames of reference moving with a velocity, v, between them. They could be the two cars from Chapter One except that this time v is relativistic. On a spacetime diagram they look like:

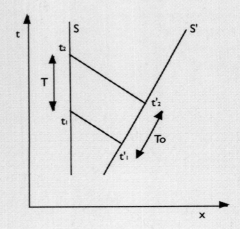

If a fixed observer in the frame of reference S', measures the time between two successive events, t'_1 and t'_2, in order to calculate the time interval as observed by someone in frame of reference S, t_1 and t_2, the Lorentz transformation must be used.

$$t_1 = \beta\left(t'_1 + \frac{vx'_A}{c^2}\right) \qquad\qquad t_2 = \beta\left(t'_1 + T_0 + \frac{vx'_A}{c^2}\right)$$

$$where \; \beta \; = \; \frac{1}{\sqrt{\left(1 - \frac{v^2}{c^2}\right)}}$$

$$x'_A = \; ordinate \; of \; the \; observer$$

In order to calculate T:
$$T = t_2 - t_1$$

which gives:
$$T = \beta T_0$$

This is the time dilation formula and shows that clocks appear to tick more slowly on moving objects by the factor β.

Figure 2.13
Relativistic radial Doppler effect

This diagram shows what would happen to a photon of light with a wavelength of 550 nanometers as the radial velocity between the source and observer increases towards the speed of light. Negative velocities are produced when the distance between the source and the observer is decreasing.

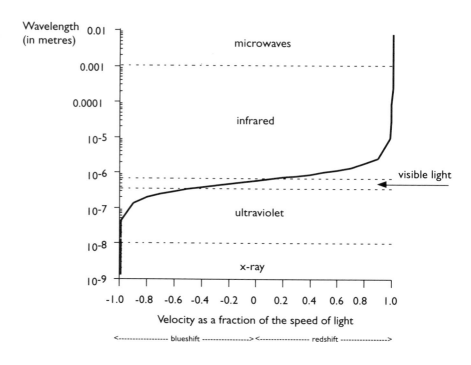

Figure 2.14
Time dilated Doppler shifts

This diagram shows a suite of curves which represent motion at various angles to the line of sight. 0° is radial motion away from the observer, 180° is radial motion towards the observer. 90° is motion in the plane of the sky and corresponds to the transverse Doppler shift. Any viewing angle greater than 90° corresponds to a component of motion towards the observer. The diagram clearly shows that if the velocity of these objects is high enough, even though they are approaching the observer, they still show a redshift because of time dilation.

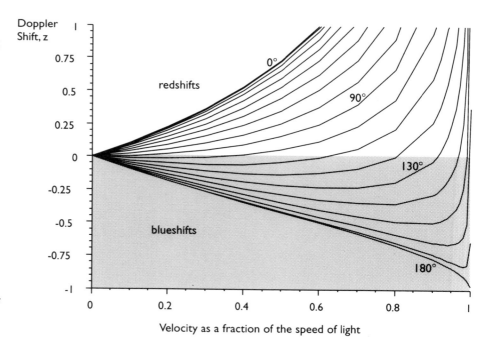

The special relativistic form of the Doppler effect, therefore, has to include the classical Doppler effect, based upon the difference in distance each wave crest has to travel, and the time dilation redshift (see *mathematical note 2.8*).

Transverse Doppler shift

The fact that the time dilation component of the relativistic Doppler shift is dependent on overall velocity, rather than just radial velocity, means that, even if the object were moving at right angles to the line of sight, a redshift would still be incurred by its emissions (see *figure 2.15*). This redshift would be totally due to time dilation and is known as the transverse Doppler shift.

The black hole candidate SS433 in the constellation of Aquila is the perfect laboratory in which to test our theories on the

Figure 2.15
Relativistic transverse Doppler shift

This diagram shows what would happen to a photon of light with a wavelength of 550 nanometers as the transverse velocity between the source and observer increases towards the speed of light. Positive velocities are east-west whilst negative velocities are west-east. Because the traverse Doppler shift is entirely due to time dilation, however it always produces a redshift.

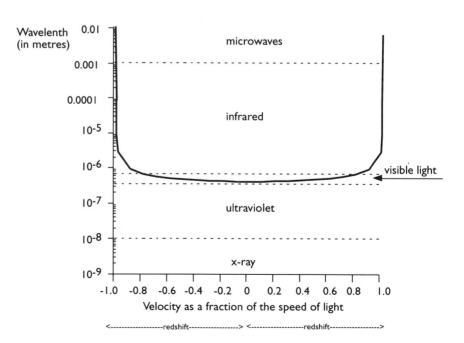

Mathematical note 2.8

The special relativistic form of the Doppler shift

In mathematical note 2.1, it was shown that the Doppler shift can be derived from:

$$T_A = T_B + \left(\frac{d'-d}{c}\right)$$

where
T_A = observed time between events from A
T_B = observed time between events from B
c = speed of light
d = distance at first event
d' = distance at second event

Also from mathematical note 2.1, the distance between events can be equated to the radial velocity, vr, in the following way:

$$d'-d = v_r T_B$$

Remembering that at relativistic velocities, time dilation takes place so:

$$T = \beta T_0$$

$$\text{where } \beta = \frac{1}{\sqrt{\left(1-\frac{v^2}{c^2}\right)}}$$

The equation to derive the Doppler shift becomes:

$$T_A = \beta T_B + \frac{\beta v_r T_B}{c}$$

Remembering from mathematical note 2.1 that:

$$\lambda = cT_A$$
$$\lambda_o = cT_B$$

where
l = observed wavelength
lo = emitted wavelength

$$\lambda = \beta\lambda_0 + \frac{\beta v_r \lambda_0}{c}$$

which simplifies to:

$$\frac{\lambda}{\lambda_0} = \beta\left(1+\frac{v_r}{c}\right)$$

Thus the special relativistic form of the Doppler shift equation is:

$$\frac{\lambda}{\lambda_0} = \frac{\left(1+\frac{v_r}{c}\right)}{\sqrt{\left(1-\frac{v^2}{c^2}\right)}}$$

Note that the denominator contains v which is the velocity of the source, whereas the numerator contains vr which is the component of v in the radial direction.

When the motion is entirely radial, i.e. v = vr, the equation becomes:

$$\frac{\lambda}{\lambda_0} = \left(\frac{1+\frac{v}{c}}{1-\frac{v}{c}}\right)^{\frac{1}{2}}$$

(see figure 2.14)

When the motion is entirely transverse, i.e. v_r=0, the equation becomes:

$$\frac{\lambda}{\lambda_0} = \frac{1}{\sqrt{\left(1-\frac{v^2}{c^2}\right)}}$$

(see figure 2.15)

transverse Doppler shift. In 1977, Bruce Stephenson and Nicholas Sanduleak classified this remarkable object as a star which possessed spectral emission lines, rather than (the more normal) spectral absorption lines. Further studies of the system soon showed that this object is far from straightforward. Its hydrogen emission lines are split into three components, one of which resides close to the laboratory wavelength of the emission. The other two roam about the spectrum with a period of 164 days. The evidence points to the fact that this system is a very close binary but the precise nature of the two stellar components is a subject of some confusion. One is thought to be very compact, either a black hole or a neutron star. Its companion has been suggested to be either a white dwarf or a blue supergiant star.

Whatever the exact identity of the components, the compact object is slowly pulling apart the companion. Since both objects are in orbit about one another, this material does not simply disappear straight into the gravitational maw of the compact object; instead, it goes into orbit around it, forming an accretion disc of hot gas. From this accretion disc, some of the material does, indeed, fall onto the compact object. The rest of it, however, for reasons which are not yet totally explainable, is expelled at right angles to the disc, causing jets to be formed. Through a mechanism which is poorly understood at present, the disc and jets also wobble. This may be because the gravity of the companion causes them to oscillate or it may be the rotation of the compact object which drags the disc around with it.

Whatever the mechanism, it is the jets which are responsible for the swirling emission lines. As they wobble, a phenomenon known as precession, they present different angles towards Earth's line of sight. At these different angles, different components of the jet material's velocity are directed along our line of sight. This results in radial Doppler wavelength shifts which alter as the jets precess. Underlying the variable wavelength shift caused by the precession is a redshift which can only be attributable to the transverse Doppler shift. This arises because the material in the jets is being flung out so fast, at about a quarter of the speed of light, that ejected material experiences a noticeable time dilation. The effect is not totally overwhelmed by the radial motion component of the Doppler shift because the jets precess about an axis which is almost at right angles to the Earth's line of sight.

Doppler boosting in active galaxies

As well as modifying the appearance of spectral lines in the central regions of active galaxies, the Doppler effect also plays a part in altering the look of the radiation being emitted by the active galactic nucleus. It is a well observed phenomenon that some active galaxies emit 'jets' of material from their centres. These jets resemble larger versions of those found around SS433. The ejected material is usually made up of sub-atomic particles, such as electrons or protons, which are accelerated to relativistic velocities and trapped by the magnetic fields present in these galactic environments. The presence of the magnetic field causes the particles to spiral along, emitting radiation. This radiation makes the jets visible because some of it is emitted in the visible region of the electromagnetic spectrum. In some cases the jets come in pairs which are directed along the rotation axis of the supposed black hole at the heart of the active galaxy. Although how these jets are produced is not yet understood, the fact that they exist is incontrovertible since a mountain of observational data has been gathered about them. Very often, however, one jet will appear brighter than the other. Sometimes the difference is dramatic with one jet blazing brilliantly whilst the other is unobservable. The reason for this lies in the Doppler effect.

Since galaxies are randomly orientated in space, these jets could be in any direction. In the majority of cases, one jet will be inclined towards the Earth whilst the other is tilted away (see plate 2.1). This means that the radiation being released by the forward tilted jet will suffer a relativistic blueshift, which increases its energy from our observational stand point. This is the phenomenon known as Doppler boosting. It must be taken into account when calculating the exact amounts of energy released

Plate 2.1
Active Galaxy M87

The core of the active galaxy M87 in the Virgo cluster shows a rotating disc of gas from which a jet of matter emanates.

(H Ford, R Harms, Z Tsvetanov, A Davidsen, G. Kriss, R Bohlin, G Hertig, L Dressel, A Kochhar, B Margon and NASA)

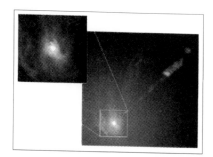

by the active galactic nucleus since the blueshifted emission gives us a biased view. In the backward tilted jet, exactly the opposite occurs, which weakens the jet by redshifting it and making it difficult, if not impossible, to observe.

3

Gravitational redshift

The laws of special relativity have shown the way an object moves affects the wavelength of radiation it either emits or observes. The special theory of relativity was a very restricted theory, however, because it only dealt with inertial frames of reference. In other words, the frames of reference had to be either stationary or moving with a constant velocity before the theory could be applied. As soon as a frame of reference changed its velocity, either by accelerating or by decelerating, the laws of special relativity no longer applied. This is why the theory was known as the special theory because it only dealt with a special kind of motion: unaccelerated. In order to determine the way in which an accelerating frame of reference 'sees' the universe, including the redshift/blueshift experienced by the radiation it comes into contact with, a better theory was required.

Einstein set about extending his theory so that it could be applied to the motion of all objects whether they were stationary, moving with a constant velocity or accelerating in some way. Thus, his new theory would apply to motion in the broadest, most general sense of the word. The complicating factor was that, unlike inertial frames of reference, accelerating frames of reference have a force acting upon them and so, for the first time, Einstein had to consider forces in his calculations. This was how the general theory of relativity was born; as an attempt to explain how the universe must appear to observers in accelerating frames of reference. Einstein realised that accelerations caused by gravitational forces could also be accounted for in his consideration of such concepts. It was one of the most important steps in modern science and has provided physicists and astronomers with the best description of gravity to date. It also predicted that light escaping from a gravitational field will have some of its energy leeched away, causing it to be redshifted.

The spacetime continuum

As has been seen in the previous chapters, special relativity tells us that both space and time are affected by movement. In fact, it was just three years after the publication of special relativity that Hermann Minkowski, a professor of physics at Zürich, proposed that space and time should be considered as linked quantities. Thus, the concept of spacetime was born. Whereas space is three dimensional (up and down, left and right, in and out), the addition of time makes spacetime four dimensional and therefore very difficult to visualise. Any occurrence in spacetime must be referenced not only by its position in space but also by the time at which it was observed to take place. Such an occurrence is called an event and its spatial and temporal co-ordinates place it somewhere in the spacetime continuum. Einstein embraced spacetime wholeheartedly and couched his ideas in its framework when he presented his general theory of relativity in 1916.

The principle of equivalence

Einstein's first step towards his general theory was to describe the principle of equivalence after considering many of the known laws of physics and embarking on a series of thought experiments. These exotically titled experiments are ones which would be either undesirable or impractical to perform in reality. Instead, they are performed in one's head and the known laws of physics are used to determine their outcome.

Einstein's famous lift experiments were just such thought experiments. He considered what would happen to a lift, containing a person and a ball, in various different circumstances. He stated that the lift was sealed in such a way that no observations of the outside world could be made. In the first circumstance the lift was taken into space, well away from any planet or star which would exert a gravitational pull upon it. A rocket motor was attached to the base of the lift and the whole thing accelerated to 9.8m/s² by ignition of the rocket. The person inside the lift would let go of the ball and watch it fall to the floor. The acceleration of the ball towards the floor would be equal, but in the opposite direction, to the acceleration of the lift.

In the second stage of the experiment, the rocket motors would be cut. The lift would continue to drift at its final velocity thus becoming an inertial frame of reference instead of an accelerated one. This time if the ball were picked up and released, it would hang motionlessly in mid-air because no forces were acting upon the lift. The occupant would feel weightless.

In the third case, the lift is brought to Earth and suspended in a lift shaft. The ball is released and we know from common sense that it would accelerate towards the floor. The famous astronomer Galileo is sometimes said to have proved, by dropping balls from the leaning Tower of Pisa, that all objects, no matter what their mass, fall to the Earth with the same acceleration. This acceleration has been measured for the Earth and is approximately 9.8 m/s². It is caused by the gravity of the Earth and is therefore known, unsurprisingly, as the acceleration due to gravity.

In the fourth and final case, the lift is cut from its suspension and allowed to fall freely down the shaft. The lift and its contents are accelerated under gravity and so this time, when the ball is released, since it is already falling downwards at 9.8m/s², it actually appears to remain stationary inside the lift.

To the person inside the lift, the first and third cases are indistinguishable, as are the second and fourth. (See figure 3.1) In the first case the force acting on the ball was being caused by the acceleration of the rocket whilst in the third case the force was gravitational in origin. Both, however, caused the ball to accelerate. Imagine further what would happen to a rocket

Figure 3.1
Einstein's thought experiments

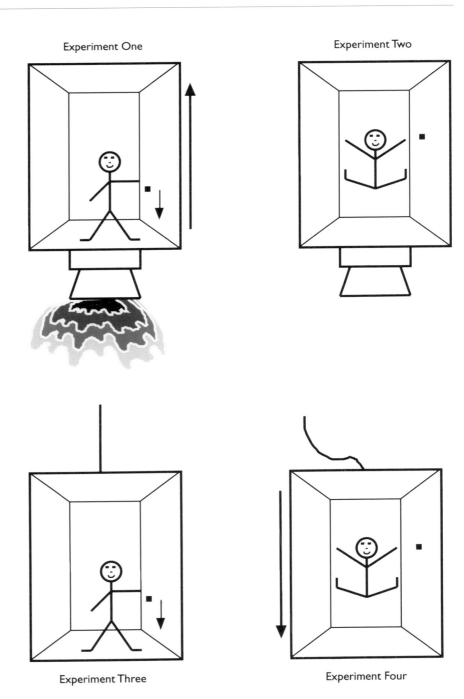

Experiment One

Experiment Two

Experiment Three

Experiment Four

accelerating away from the Earth at 9.8m/s². Since the downward pull of gravity would try to accelerate it in the opposite direction at exactly the same rate, the rocket would get nowhere.

To place the conclusions of these thought experiments in a somewhat more concise way: Einstein stated that accelerations and gravitational fields are indistinguishable from one another. So indistinguishable, in fact, that a correctly chosen and applied acceleration can counteract the effects of a gravitational field and vice versa. In essence, this is the principle of equivalence and shows why general relativity, which sought to describe an accelerating frame of reference's view of the universe, also describes gravity.

It was easily seen in figure 2.1 that motion produces a Doppler shift which can be either a redshift or a blueshift depending upon the relative motion between the source and the observer. At any specific instant in time, an accelerating object can be though of as having a constant velocity. For instance, twenty seconds after it has started moving, an object may be travelling at 10m/s and producing a corresponding Doppler shift. Ten seconds later, that object may now be travelling at 15m/s and hence producing a Doppler shift which corresponds to this new velocity. So, as the object has accelerated, the Doppler shift it creates has changed (see figure 3.2).

Since the principle of equivalence states that the effects of an accelerating frame of reference are indistinguishable from that of a gravitational field, it follows that a gravitational field must also be able to impart redshifts and blueshifts on radiation.

The gravitational wavelength shift

To illustrate the way in which the wavelength of radiation is altered by a gravitational field, imagine the lift from the fourth case of the thought experiment, i.e. falling freely under gravity towards the Earth. A beam of light is released upwards from the floor of the lift and strikes a detector on its roof. Because the lift and the light beam are both being accelerated downwards due to gravity, the detector measures the

Figure 3.2
The accelerated Doppler effect

Positions a to e correspond to the position of the source when wave crests A to E were released. The time interval between successive wave crest emissions is a constant. Hence, the source is accelerating because the distances between emission positions are becoming greater. The wavelength shift, as indicated by the distance between sucessive wave crests becomes more pronounced the faster the source moves.

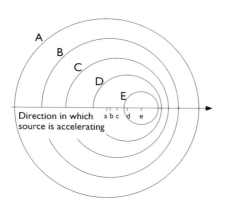

Mathematical note 3.1

The gravitational redshift

If we consider the free falling lift in Einstein's fourth thought experiment. Light of frequency f_e is released from the base of the lift just as the cable is cut and the lift enters free-fall. The light hits a detector in the ceiling of the lift after a time:

$$t = \frac{h}{c}$$

where
h = the height of the lift
c = the speed of light.

Since the detector is in the same frame of reference as the source from which the light was emitted, it detects light of frequency f_e. A detector outside the lift, however, would appear to be in motion relative to the lift. If it were possible to arrange the detector so that it would detect the radiation at the moment it passed through the lift ceiling, the lift would be travelling downwards with a speed:

$$u = gt = \frac{gh}{c}$$

where
g = gravitational acceleration of the Earth $(9.8m/s^2)$

This can now be used to give the redshift, z, of the light as viewed by the outside detector:

Remembering from mathematical note 2.1,

$$z = \frac{u}{c}$$

Substituting for u gives:

$$z = \frac{gh}{c^2}$$

The quantity gh is the change in gravitational potential experienced by the light. The difference in the gravitational potential will be equal to the energy expended by the photon in climbing through the distance h. This quantity of energy is often referred to as the work done.

wavelength of radiation to be the same as that with which it was emitted. This result should come as no surprise since both the emitter and the detector are in the same frame of reference. If the radiation is allowed to pass out of the roof, however, and strike a detector which is not being accelerated by gravity, the radiation will appear to have its wavelength redshifted *(see mathematical note 3.1)*. This is because the lift and the light source appear to be moving away from the detector and we have seen in Chapter Two that moving sources incur wavelength shifts in their radiation. For radiation emitted from the top of the free falling lift downwards, a detector below the lift would see the emitted radiation being blue shifted.

The explanation chosen is an implicit form of the principle of equivalence since, although we are discussing wavelength shifts caused by radiation either falling into or climbing out of gravitational fields, we have described them in terms of the lift's accelerated motion. We have thus proved that the action of gravitational fields is indeed equivalent to that of accelerating frames of reference.

Another way to think about the gravitational redshift is to imagine rockets leaving the Earth. No one would dispute that a rocket needs to expend energy in order to escape the gravitational attraction of the Earth. It does this by burning fuel in its engines to produce an acceleration which firstly counteracts the force of the Earth's gravity and then pushes the rocket away from the Earth. Common sense dictates that the amount of energy needed to accelerate the rocket is dependent upon its mass. The more massive the rocket the more energy will be required in order to accelerate it sufficiently.

As a consequence of special relativity Einstein had shown that energy and mass are interchangeable quantities; mass is the embodiment of energy. To use a financial analogy, mass is like a business' fixed assets whilst energy is like its fluid assets. Einstein's introduction of the concept that light was composed of photons led to the idea that each of the particle-like bundles of energy would behave as if they had small masses. Those masses would be dependent upon the energy contained by the individual photons (see mathematical note 3.2). Thus to escape a gravitational field, any photon must expend some of its energy in order to overcome the gravitational force acting on it. When a photon loses energy, as was stated in Chapter One, it suffers an increase in its wavelength, i.e. it is redshifted. Hence, any

---

Mathematical note 3.2

The 'mass' of a photon

We have already established in mathematical note 1.5 that the energy of a photon is:

$$E = hf$$

where
E = energy of a photon
h = Planck's constant
f = frequency of light.

Einstein's famous equation which describes the interchangeability of mass and energy is:

$$E = mc^2$$

where
E = energy
m = mass
c = speed of light.

Combining the two equations gives the 'mass' of a photon as:

$$m = \frac{hf}{c^2}$$

photon which is leaving a gravitational field will suffer a redshift, the magnitude of which is principally determined by the strength of the gravitational field and, to a lesser extent, the energy of the photon. Conversely, any photon falling into a gravitational field will suffer a gravitational blueshift, the magnitude of which will be equal but opposite to the gravitational redshift.

Measuring the gravitational redshift

The ability of general relativity to predict this hitherto unsuspected gravitational redshift meant that it was a natural target for the experimentalists and observers. At first they levelled their instruments at the Sun but, despite the Sun's great mass, the redshift it produced was still too small for the technology of the day to measure conclusively.

In 1925, however, Adams successfully measured the gravitational redshift from a star known as Sirius B. This star is the tiny companion to Sirius, the brightest star in the night sky. It is a class of object known as a white dwarf which is made up of the highly compact remains of a dead star. Although these stars are very small, approximately the diameter of the Earth, they are extremely dense and possess very strong gravitational fields. These fields are so strong that the gravitational redshift is thirty times greater than that produced by the gravitational field of the Sun. In fact, it was thirty seven more years before technology had advanced sufficiently for J.E. Blamont and F. Roddier, of the Meudon Observatory in France, to be able to measure successfully the gravitational redshift of the Sun.

Ironically, the measurement of the Sun's gravitational redshift was predated by a measurement of the Earth's much weaker redshift in 1960. This extraordinarily precise measurement was made possible only by the discovery of the Mössbauer effect. In 1958, German physicist Rudolf L. Mössbauer had explained that identical atomic nuclei must have identical electron energy levels around them. Thus a photon emitted by one nucleus would be absorbed by another. In this event, the nuclei were said to be in resonance. If anything even slightly changes the energy of the emitted photon it will not be absorbed by the receiving nucleus. The technique is so sensitive that, in theory, it could detect a Doppler shift produced by a velocity of just one inch every thousand years!

This technique, in conjunction with a 23 metre-high tower at Harvard, was used by Pound and Rebka in 1960 to measure the wavelength shift of a 14.4 keV gamma ray emitted by the radioactive isotope ^{57}Fe. The Mössbauer effect allowed them to exactly measure the minuscule wavelength shift predicted by general relativity.

Since then, other experiments have also allowed measurements to take place. Most other measurements of the gravitational redshift rely on devices known as atomic clocks. In order to understand these exotic devices, the interplay between the emission of radiation and the passage of time must be explained. In Chapter Two, the way in which radiation is redshifted when emitted by an object travelling away from an observer was explained. That chapter also introduced the concept of time dilation which states that, to outside observers, time appears to travel at a slower rate on a moving object. Since a slowing down in the passage of time corresponds to a redshift being imparted to the radiation, is it possible to explain the gravitational redshift in terms of the passage of time?

The explanation involves the concept of atoms being used as clocks. Imagine an atom which is continuously releasing radiation of a known wavelength. Each time a wave crest is released think of this as a 'tick' on the atomic clock. If the atom is accelerated to relativistic velocities, time dilation will affect it. The passage of time will appear to slow down for outside observers and the time between each 'tick' will appear to lengthen. It is this phenomenon which gives rise to the transverse Doppler shift as explained in Chapter Two.

A similar time dilation explanation can be applied to gravitational redshift. Since the wavelength of radiation leaving the gravitational influence of an object is redshifted, this would suggest that time passes at a slower rate in the proximity of that object. In order to test this assertion, some from of naturally occurring variation has to be produced and measured. This method of marking the passage of time is really just a high technology extension of using quartz crystals in wrist watches. The quartz crystals naturally oscillate at a frequency which is used to display the correct time on the watch. In order to measure the minute effects of the gravitational redshift, however, these variations must be ultra stable. These type of 'clocks' are given exotic sounding names which describe how they produce such stable oscillations. These devices include the hydrogen maser clocks, superconducting-cavity stabilised oscillator (SCSO) clocks and clocks containing cryogenically cooled monocrystals of dielectric materials such as silicon and sapphire! The key to obtaining such precise oscillation frequencies is to average the time between each individual oscillation which occurs in a period of time. That period is usually between ten and one hundred seconds.

In 1976 a hydrogen-maser clock was launched by NASA to an altitude of 10,000km in order for its frequency to be compared with an identical hydrogen-maser clock which remained safely on the ground. After the Doppler effect, caused by the motion of the rocket, had been eliminated, a tiny change in the frequency was still found. When it was compared to the ground-based clock it was found that the frequency of the clock's 'ticks' had increased. This was exactly as the theory predicted. Since the presence of a gravitational field was thought to slow the passage of time, the passage of time should speed up when away from the gravitational influence of the

Earth. Indeed, as the clock got further and further away from the surface of our planet so it ticked faster and faster as the gravitational influence on it was reduced.

The confirmation of the gravitational redshift has been crucial in the development of navigational satellites. For instance, the satellite NAV-STAR 2, part of the U.S. Global Positioning System, has to take the time dilating effects of both special and general relativity into account in order to return accurate data to Earth from its atomic clock. It orbits at a heights of some 20,000km and, because of its significant velocity with respect to its ground stations, special relativity predicts that time will appear to slow down causing a redshift when it transmits data home. Since the satellite is so far from the surface of the Earth, however, general relativity predicts that the passage of time will speed up causing the atomic clock data to be blueshifted! The two effects would cancel each other out if the satellite were in an orbit one and a half times the radius of the Earth but, since the orbit is actually just over four times the Earth radius, the atomic clock runs faster by about 38.5 microseconds every day. If this went uncorrected it would introduce errors in the tracking data of some 11 kilometres per day.

So far our discussions of the gravitational redshift have indicated that these variations are very small but this is only true in weak gravitational fields such as those found surrounding planets like the Earth and stars like the Sun. There are regions of the universe in which the gravitational redshift becomes so powerful that it redshifts radiation out of universal existence!

Black holes

These astronomical objects hold a place of special importance in the twentieth century. They are so outrageous that it is difficult to believe any rational scientific theory could have predicted their existence. The abstract notion of a type of object which would later be termed a black hole had its birth in the works of Englishman John Mitchell and Frenchman Pierre Simon de Laplace in the later years of the eighteenth century. Both men were conversant with Newton's law of motion and his theory of gravitation. In his work Newton had presented the notion of escape velocity. He knew that any object with mass also produced gravity. He reasoned that in order for another object to leave its gravitational influence, the departing object would have to accelerate to a certain velocity which would overwhelm the force of gravity acting upon it. That velocity would obviously depend upon the mass of the gravitating body, the more massive it was, the greater the velocity would need to be in order to escape. Thus, the concept of escape velocity was born. Mitchell and Laplace both wondered what would happen if an object was so massive that its escape velocity equalled the speed of light. They reasoned that any such objects would appear black since no light could escape from them but beyond this conclusion, work was foiled by the inadequacies of Newton's gravitational theory. With the advent of general relativity, however, physicists were suddenly empowered with a means of theorising about such fascinating objects.

To remain in Newton's framework for one last analogy, it can be deduced very easily that every object in the universe could be turned into a black hole. A star, a planet, a building, even a person, could become a black hole if the mass they possessed were to be squeezed into a sufficiently tiny volume. Usually it is conceived that this volume is spherical and can be defined by something known as the Schwarzschild radius (see *mathematical note 3.3*). Figure 3.3 gives a table of Schwarzschild Radii for various objects.

But what in nature can cause such squeezing? If we look into our galaxy the only objects which are in the process of collapse are vast clouds of molecular gas. We know from extensive observations that, when these gas clouds collapse, they form

Mathematical note 3.3

The Schwarzschild radius

The gravitational force between two objects with mass is given by:

$$F = \frac{GM_1 M_2}{r^2}$$

where
F = force of gravity
M_1 = mass of object one
M_2 = mass of object two
r = distance between the centres of M_1 and M_2
G = gravitational constant

This is shown diagrammatically as:

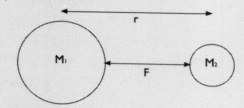

Imagine that the mass M_2 is very small, so small, in fact, that it can sit on the surface of M_1. This would be the situation if, for example, M_1 were a planet and M_2 were a marble. r is then the radius of the planet. Imagine next that some way was found to gently loft M_2 to an infinite distance away from M_1. In order to discover how much energy M_2 had expended in this effort, the gravitational force equation would need to be integrated from M_2's starting point, r to its finishing point, ∞.

$$E = \int_r^\infty \frac{GM_1 M_2}{r^2} dr$$

where
E = energy expended
G = gravitational constant
M_1 = mass of planet
M_2 = mass of marble
r = radius of M_1

Performing this integration gives:

$$E = \frac{GM_1 M_2}{r}$$

E, the energy expended by M_2 has been turned into movement and so has become kinetic energy. This is given by the equation:

$$E = \frac{1}{2} M_2 v^2$$

where
E = kinetic energy
M_2 = mass of marble
v = velocity of marble

equating the two gives:

$$\frac{1}{2} M_2 v^2 = \frac{GM_1 M_2}{r}$$

which can be rearranged to:

$$v = \sqrt{\frac{2GM_1}{r}}$$

v is known as the escape velocity since it is the velocity needed to get the marble out of the gravitational influence of M_1 and all the way to infinity. It is usually denoted v_e. It is worth noting here that the mass of the escaping object, in this case the marble, has cancelled out of the equation. Hence, the escape velocity is the same no matter how massive the object trying to escape.

If we set v_e = speed of light and rearrange to give us the radius, we obtain the Schwarzschild radius of a black hole with mass M_1:

$$r_s = \frac{2GM_1}{c^2}$$

where
r_s = Schwarzschild radius
M_1 = mass
G = gravitational constant
c = speed of light

Figure 3.3
Schwarzschild radii

This table shows the Schwarzschild radii for various celestial bodies. In reality the only objects likely to reach their Schwarzschild radii would be the final two on the list. In the case of a supergiant star, it would be only the core which became a black hole during a supernova explosion. The core would possess a mass between those of the Sun and the supergiant star in the table.

The other astrophysical situation in which black holes are thought to form is when matter streams into the centre of the galaxy. Here the required density of material is reached naturally as material is gathered up by the galaxy's pull of gravity. It may be that all galaxies have black holes at their centres. In active galaxies, so much energy is produced in their cores, that a black hole swallowing matter is the best explanation for how it produces so much power.

Object	Mass	Schwarzschild Radius
The Moon	7.34×10^{22} kg	0.109 millimetres
The Earth	5.97×10^{24} kg	8.85 millimetres
Jupiter	1.899×10^{27} kg	2.8 metres
The Sun	1.99×10^{30} kg	2.95 kilometres
A Supergiant Star	4×10^{31} kg	59.3 kilometres
Central Regions of an Active Galaxy	2×10^{39} kg	1×10^9 kilometres

stars. By studying this process astronomers have produced a detailed theory to explain why the collapse to black holes does not proceed unabated. As the cloud material becomes more and more dense, so the very centre of the condensation is squeezed to greater and greater densities. The temperature of this core region is raised by this process so much that the atoms gain a tremendous amount of movement energy. This, combined with their high proximity, means that when they collide the process of nuclear fusion takes place. Indeed, it is the ignition of nuclear fusion which signals the birth of a star. The copious amounts of energy released by the fusion process 'support' the star and stop it from collapsing any further. Although, at first, this would appear to ruin our idea of how black holes form, as will be seen, nuclear fusion is only a temporary respite from the inexorable collapse into cosmic oblivion!

All the stars in the night sky, indeed within the universe, must have gone through this formation process. Stars comes in all sizes, the high mass stars are very hot whilst the low mass stars are cool. The mass of a star is dependent upon the speed with which its natal cloud collapsed. The more material accumulated before the onset of nuclear fusion, the more massive the final star. Whatever the size, the driving source of a star's energy is nuclear fusion but the mass of the star defines how long it can survive. More massive stars burn themselves out in shorter time periods whilst less massive stars can persist for longer. Stars such as the Sun can survive for about nine billion years. The mass of a star also signals just how it will conclude its existence.

Low mass stars, ie those under the approximate limit of five times the mass of the Sun, will gradually shed their outer layers of gas as fusion stops and the remains of the star will begin to collapse once again. Although this looks hopeful for the creation of a black hole, the collapse soon stops because the electrons in the collapsing remnant resist being pushed too close to their respective atomic nuclei. They exert a resistive force, known as electron degeneracy pressure, which supports the stellar remnant. The object produced is known as a white dwarf and astronomers know of hundreds, scattered around our galaxy.

High mass stars suffer much more spectacular fates. These stars are capable of supporting a more varied range of nuclear fusion reactions within their centres. This is because the pressure on those central regions, caused by the weight of the surrounding material, is so large. It is in the hearts of these massive stars that all the lighter chemical elements up to and including iron are synthesized. Iron is the twenty sixth element in the periodic table and marks a fundamental dividing line in the way elements take part in nuclear reactions. Up to this point all chemical elements have been able to fuse into heavier elements and convert some mass into energy. From iron onwards, in order for fusion to take place, the reaction must be fed additional energy. No matter how much the central region of the star is squeezed, additional energy in not forthcoming and the inert iron core collapses. The core shrinks to a fraction of its former volume in the blink of an eye. In fact, it happens so fast that the surrounding star continues to fuse other, lighter, elements and remains completely unaware of the drama taking place in its centre. However, as the seconds tick by, in the same way that a disused tower block collapses when the demolition team blows out its foundations, so the star gradually falls in on itself.

The shock wave, produced by the stellar material striking the collapsed core, gradually propagates outwards through the star, causing it to explode in a phenomenon called a supernova. The collapsed core is much denser than a white dwarf since the electrons have actually been forced into the atomic nuclei where they fuse with the protons and form neutrons. The resultant objects are known to astronomers as neutron stars. In theory, the violent activity around a collapsed core should sometimes be able to provide enough mass to squeeze the neutron star past the Schwarzschild limit and produce a black hole.

Having now identified where astronomers expect black holes to be formed, general relativity can next be invoked to help tell us what to expect from them. This chapter has introduced the idea of the spacetime continuum. The way in which gravity is explained in this context is to state that it is produced by curvature of the spacetime continuum. To make this idea clearer, imagine a rubber table top. If a heavy ball were placed upon it, the rubber would sag around it. If a marble were now placed anywhere near the heavy ball, the curvature of the rubber would cause the marble to roll towards the heavy ball and touch it. This is a good analogy which should be kept in mind when thinking about how mass produces gravity. It also gives rise to the term gravitational well which refers to a massive object's distortion of the surrounding spacetime continuum.

With this analogy in mind, the formation of a black hole can be understood in the following way. Imagine that the heavy object in the centre of the table is the inert core of a star undergoing gravitational collapse. The smaller it gets, the denser it gets, the greater the gravitational field it creates and hence the more the rubber table top sags. In our analogy, the greater the gravitational force becomes, the deeper the gravitational well becomes and the greater the distortion in the rubber table top.

Eventually the density of the central object becomes so great that the rubber will be stretched downwards into an infinitely deep 'hole'. As the object collapses past the Schwarzschild radius, so it becomes a black hole. Essentially, this means that it has lost its ability to communicate with the surrounding universe since nothing, not even light, can escape from it.

The Schwarzschild radius delineates one of the most important type of boundaries in the universe. We have already introduced the concept of an event which was a happenstance in space and time. Such events, which could be anything from doing your housework to the explosion of a star, are observed by others because radiation is released or reflected and the event can be observed. Since radiation cannot escape from the object once the Schwarzschild radius has been crossed, this marks the event horizon. We can never know what takes place inside the event horizon of a black hole since the known laws of physics are inadequate to describe these regions.

We have talked about the photons of light being unable to escape from the strong gravitational fields of black holes. In our initial description of gravitational redshift we have thought of the photons trying to struggle upwards and out of the gravitational well. This explanation may give the impression that the photons get as far away from the black hole as they possibly can, before succumbing to the overwhelming pull of gravity and sinking back down again. As we have seen, a more physically realistic way of thinking of the situation is available to us thanks to the concepts of general relativity and gravitational time dilation.

As the gravitational field becomes stronger and stronger, so does the effect of the time dilation and the radiation becomes more and more redshifted. It is important to remember that the photons of radiation still leave the surface of the contracting object at the speed of light but as the object gets smaller and smaller, so the emitted light becomes redder and redder. By the

Figure 3.4
Redshifting radiation

As the stellar core collapses, so the gravitational field around it becomes stronger. This stretches the wavelength of emitted radiation causing it to become more and more redshifted. As the object reaches its Schwarzschild radius, the redshift becomes infinite and the emission is no longer detected. The object has become a black hole.

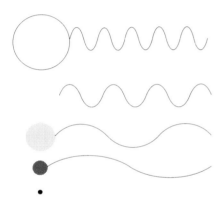

time the object reaches the dimensions of the Schwarzschild radius, the redshift has becomes infinite (see figure 3.4). In other words, the electromagnetic waves are stretched so much that their wavelengths become infinite. An infinite wavelength corresponds to the wave carrying zero energy and so radiation stops being emitted and they appear totally black. This fact, combined with the object's distorting effects on the spacetime continuum is where the name black hole derives. Also, since the redshift has become infinitely large at the event horizon, this must mean that, at that boundary, time has ceased to flow!

Almost unbelievably, it may still be possible to detect these objects. If a black hole were to exist within a binary star system, such as the ones which produced spectroscopic binaries discussed in Chapter Two, then it may be possible to infer its existence at a certain point in the binary's evolutionary life cycle. This point occurs when the black hole's companion begins to change the nature of the nuclear fusion taking place in its core. The introduction of helium fusion bloats the star into a red giant which will cause gaseous material to cross into the gravitational influence of the black hole. The material will fall down the gravitational well and into the black hole like water spiralling down a plug hole. When this happens, the gas travels so fast that friction with neighbouring molecules causes it to be heated, which in turn causes it to emit radiation. The radiation is in the form of x-rays and can be detected from Earth with suitable equipment. Several such x-ray sources are known to exist within binary star systems. Probably the most famous of these is located in the constellation of Cygnus. It is known as Cygnus X-1 and is thought by many to be one of the best black hole candidates in the galaxy (see figure 3.5).

Black holes in active galaxies

The idea of detecting black holes by looking for the x-rays given off from the gas in their accretion discs can be extended to the

Figure 3.5
Cygnus X-1

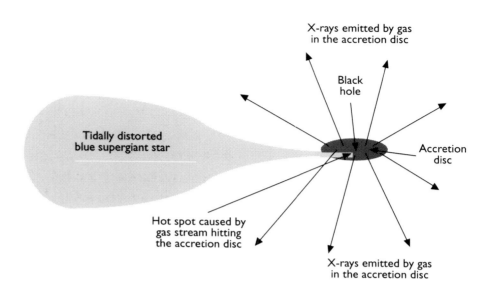

supermassive black holes thought to exist at the hearts of active galaxies. Active galaxies were introduced in Chapter Two and we have seen in this chapter that the black holes, which lie at their centres, are very massive indeed. Astronomers should be able to detect the gravitational redshift of the radiation given off from the very inner regions of the accretion disc, just before the gas swirls into the black hole. Observations of the active galaxy, MCG-6-30-15, a Seyfert galaxy in the constellation Centaurus, show not only this effect but several others as well. In fact, MCG-6-30-15 enters the galaxy 'hall of fame' by being the only object in the universe so far observed to display all three types of the redshift phenomenon simultaneously in its spectrum!

In 1994, a team of twelve astronomers, led by Yasuo Tanaka, used the Advanced Satellite for Cosmology and Astrophysics (ASCA) and observed a peculiarly shaped x-ray emission line in the centre of the galaxy. Having identified it as being the Kα line emitted by iron atoms, they then set about understanding why it was skewed towards the weaker x-ray energies. Using the best estimates for the mass of the central black hole in MCG-6-30-15, the astronomers still found a residual redshift in the data even after they calculated the expected gravitational redshift.

They explained it in the following way. When the gas in the accretion disc is so close to the black hole's event horizon that it can be gravitationally redshifted, the intense gravitational field of the black hole must also be pulling the gas around in orbits at relativistic velocities. Usually this would produce a redshifted 'wing' and a blueshifted 'wing' as material speeds towards the Earth on one side of the black hole and away from it on the other. The emission from MCG-6-30-15 only shows a redshifted wing, however. This means that the accretion disc must be almost at right angles to us and the redshift is produced by the transverse Doppler shift.

Finally, upon leaving the galaxy altogether, the entire radiation output of the galaxy, which includes the accretion disc's emission line, is cosmologically redshifted by the expanding universe as will be explained in the next chapter.

The shadow of creation

The gravitational redshift is also responsible for one of the most celebrated astronomical observations of the last decade. It has been known since 1965 that the universe is bathed in a perpetual glow of microwave radiation, called the cosmic microwave background radiation. It is the radiation left over from the creation of matter in the early universe and has been redshifted, as a whole, by the cosmological expansion of the universe as will be explained in the next chapter. Superimposed upon this, however, is a gravitational redshift which allows astronomers to investigate the distribution of matter in the early universe.

The radiation generated in the early universe was of the continuous spectrum type. As noted in Chapter Two, continuous spectra are emitted from hot bodies. The number of photons at each wavelength in a spectrum defines its energy distribution. In turn, that energy distribution defines how hot the emitting body was when the radiation was released.

At the point when the cosmic microwave background radiation was released into space, about 300,000 years after the big bang, the universe had the same temperature throughout its volume. The background radiation should therefore show the same energy distribution in whatever direction ones cares to look into space. For almost thirty years this is the way it seemed, until 1992, when an exciting announcement was made. An American satellite named the Cosmic Microwave Background Explorer (COBE for short) had discovered that certain areas of the universe had been at slightly different temperatures when the microwave background radiation was released. True the variations were tiny, in fact, they were less than one ten thousandth of a degree centigrade but these deviations could be showing only one thing: the primordial distribution of matter out of which grew today's universal structure.

The reason for reaching this conclusion is actually straightforward. It relies upon the assumption that the variations in temperature arise because the radiation has been gravitationally redshifted. Regions of space, where the density of matter was higher than average, imparted a stronger redshift than other areas, where the density was less than average. As usual, these differences in redshift manifested themselves as differences in wavelength of the background radiation. When these differences were related to emitting regions, dense areas appeared cool whilst the sparse areas appear hotter.

The results tell us that 300,000 years after the beginning of the universe, it had already developed a clumpy structure to its matter content. It is in these clumpy regions that the galaxies, stars, planets and, eventually we ourselves then formed.

4

Cosmological redshift

For thousands of years, mankind had contented itself with staring up at the heavens using just naked eyes. That changed in 1609 when Galileo adapted the design of a Dutch spectacle maker, Hans Lippershey, and made a telescope. At the time, Galileo was studying at the university of Padua. From descriptions of Lippershey's designs, Galileo constructed a series of telescopes, improving their magnification from nine to thirty times in the space of a year. Galileo's major discoveries included that the Moon possessed mountains and craters, the Milky Way was composed of hundreds of stars and that four satellites always orbited Jupiter.

Galileo's study of astronomy was cut short by the Roman Catholic inquisition. Even after successfully defending himself against a heresy charge he was still forced to recant his observational conclusions. He was also forbidden to study astronomy and placed under house arrest for the remainder of his life.

Despite this ignoble end to his illustrious astronomical career, Galileo had opened the observational flood gates. Many followed in his footsteps developing improved telescopes. Isaac Newton designed a type of telescope which is still in use today and is known eponymously as the Newtonian. As observations of the cosmos became ever more detailed and accurate, it became obvious that the heavens contained other objects apart from simple stars. To the telescopes of the day, these objects were often nothing more than indistinct fuzzy patches of light. As a result, they became known by the generic term of nebulae which is the Latin word for clouds.

In a paper published in 1755, German philosopher and cosmologist Immanuel Kant displayed remarkable insight and expounded his ideas that the Milky Way was no different from any other nebulae, except that the Sun was immersed within it. The other nebulae were fundamentally the same type of objects but much further away.

The first list of nebulae was compiled by French astronomer Charles Messier and published in 1774 with supplements in 1780 and 1781. Messier was an accomplished comet hunter who found the nebulae an annoyance because, superficially, they looked like comets and caused him many false alarms. In an attempt to avoid this, Messier compiled a list of nebulae using his own discoveries and those shared with him by four other astronomers (among them Edmond Halley).

A contemporary of Messier's was the German-born astronomer Sir William Herschel. He came to live in England and finally built his own telescope whilst he was in his thirties. Nine years after the publication of Messier's famous catalogue, Herschel, at the age of forty five, began systematically sweeping the sky in order to discover more examples of nebulae. With the help of his sister, Caroline, and his son, John, Herschel discovered over 2,500 nebulae in a study lasting twenty years. He presented his results to the Royal Society in a series of three catalogues between the years 1786 and 1802.

During his time surveying the heavens, he came to suspect that the nebulae could be split into two fundamentally different types. Conscious of the fact that the Milky Way looked like a fuzzy patch of light until it was resolved into stars by a telescope, Herschel suggested that some nebulae were very distant star systems. Interestingly enough, he correctly concluded that the Orion nebula was a 'true' nebula and therefore relatively close by. Later in life, however, Herschel began to doubt his original ideas about distant star systems and, in 1791, abandoned his earlier assertions.

Following his father's death in 1822, John Herschel continued the work. He eventually took his father's telescope to the Cape of Good Hope in order to extend the survey to the southern hemisphere. In 1864 he published the General Catalogue of Nebulae which contained 5,079 objects, 4,630 of which had been discovered by either him or his father.

The next great piece in the understanding nebulae puzzle was uncovered by the 3rd Earl of Rosse at the Birr Castle Observatory in 1845. He built what, for its day, was a gargantuan 72 inch reflecting telescope and discovered that many nebulae were spiral in form. For many years the Birr Castle reflector was the only telescope capable of showing the spiral form of nebulae and as a result the castle became something of an astronomical mecca. During the years between 1874 and 1878 Danish astronomer J.L.E. Dreyer was assistant at the castle observatory. During his time at Birr he collected much of the information he later used in creating the New General Catalogue of Clusters and Nebulae. It was originally published in 1888 with supplements in 1895 and 1908. It is still in wide-spread use by professional astronomers as it contains over 13,000 objects.

By the turn of the twentieth century debate concerning the nature of the so-called spiral nebulae was really hotting up. Despite the fact that some of the most esteemed astronomers had supported the extragalactic hypothesis for these objects, the weight of astronomical opinion rejected it. Partly, this was due to ignorance about the extent of our own galaxy, the Milky Way. That gap in our knowledge was filled in 1917 by Harlow Shapley who correctly analysed the distribution of ninety three globular star clusters and discovered that the Galaxy was vast: at least 100,000 light years across (see figure 4.1).

By now it had become very obvious that nebulae could be split into two distinctly different groups. The first are almost always found close to the Milky Way and are amorphous in shape. The second type are dotted all over the sky and are either elliptical or spiral in shape. The key to understanding each type of object was wrapped up in its spectra. The first type of nebula displayed spectra which contained bright emission lines. These, as we saw in Chapter One, are produced by hot gas and so the term nebula really did fit these objects.

Figure 4.1
Shapley's analysis of the extent of the Milky Way

Harlow Shapley measured the distances to many globular clusters and found that they exist in a very non-uniform pattern when centred on the Earth. By assuming that they produced a radial pattern around the Galaxy's centre of mass, he showed that the Sun was situated many, many light years away from the centre of the Milky Way.

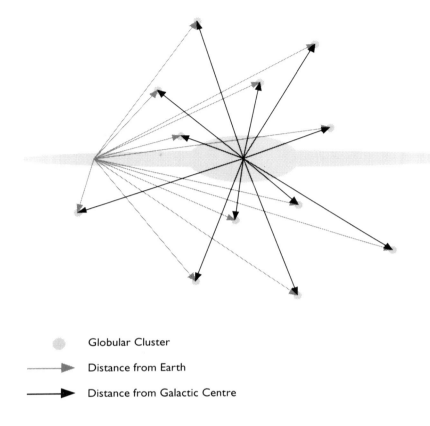

Globular Cluster

Distance from Earth

Distance from Galactic Centre

Photographing spectra was at the cutting edge of astronomical technology in the early decades of the 1900s, due to the fact that it was incredibly difficult to accomplish. The American astronomer, Percival Lowell, was another famous name to drive this subject forward. To many he will always be remembered because of his fascination with the idea of life on Mars. Indeed, in order to study the red planet, he established an observatory at Flagstaff, Arizona. As well as believing he could see canals on Mars, Lowell was also of the opinion that spiral nebulae were solar systems in the process of formation and therefore part of our own Milky Way.

In an effort to prove this assertion, he instructed one of the observatory's young astronomers to photograph a spectrum from a spiral nebula. It was Lowell's hope that it would show the spectral lines of familiar elements found in solar systems. The young man charged with this endeavour was Vesto M. Slipher. It was no easy task, either, since the 24 inch refractor at Flagstaff was not really suited to the job. A typical exposure to capture one single spectra took anywhere from between twenty to forty hours! Nevertheless, Slipher obtained his first spectra of a spiral nebula in 1912 and it became immediately obvious that something unexpected was going on. His target was, unsurprisingly, the brightest of the spiral nebulae: M31, the Andromeda nebula. The spectrum gave no hint at all that planets were present. In fact, the spectrum showed absorption lines, confirming Huggins' visual observation of the spectrum in the mid 1800s. This immediately proved that the Andromeda nebula was fundamentally different from the amorphous nebulae because those objects displayed spectral emission lines. A further peculiarity of the Andromeda data was that the absorption lines were blueshifted so much that they indicated the object was travelling towards Earth at a substantial velocity. By assuming that the blueshift was a Doppler effect, the velocity was calculated to be in the region of 300km/s. This was a much greater velocity than was usual for stars and spurred Slipher on, to investigate other spiral nebulae.

In 1914, Slipher presented his latest findings at the American Astronomical Society meeting at Northwestern University, Evanston. He showed that, in all of his observations of spiral nebulae, he always obtained spectra which contained absorption lines and displayed large Doppler shifts. In most cases, unlike M31, those Doppler shifts were towards the red end of the spectrum, indicating that the spiral nebulae were moving away from the Earth. Slipher also found that the velocities associated with these spiral nebulae were huge. Some made his Andromeda figure of 300km/s look sedate!

Hubble's work on galaxies

As fate would have it, present in the audience for Slipher's talk was a new graduate student named Edwin P. Hubble who was attending the meeting on the suggestion of his doctoral supervisor, Edwin B. Frost of the Yerkes Observatory, Wisconsin. Perhaps captured by the mystery of what Slipher's work was showing, Hubble set about photographing and classifying spiral nebulae for his thesis. Finishing this work and following a brief spell in the American Army during World War I, Hubble continued his research at the Mount Wilson Observatory in California.

In 1924 Hubble finally ended the, sometimes bitter, debate about the nature of spiral nebulae. Using the 200-inch Hale telescope, then the largest in the world, he correctly identified Cepheid variable stars in spiral nebulae, M31 and M33.

Cepheid variables are unstable stars which have left the stable hydrogen-fusing stage of their existence. They have begun the internal changes which will result in their reaching the helium fusion stage and becoming red giant stars. During this brief transitional phase of their lives, a Cepheid will change its luminosity in a repeating cycle. The time it takes to complete this cycle is linked to the star's luminosity: the brighter the star, the longer the cycles takes.

In principle, Hubble followed three simple steps. He measured the apparent brightness of a Cepheid variable star in a spiral nebulae. He also measured the period of its variation and used that to calculate how bright the star would appear if it were close by. Finally, he calculated how far away the star, and hence the spiral nebulae in which it was found, had to be in order for it to appear as dim as it did (see *mathematical note 4.1*).

His work showed that both M31 and M33 are about 930,000 light years distant. Shapley's work had shown the Milky Way in

Mathematical note 4.1

The distance modulus

The eye responds to brightness in a logarithmic way. How bright a star appears, a quantity known as its magnitude, depends upon how much radiation, known as the flux, arrives at the Earth and is observed. The magnitude of two stars can be compared using Pogson's equation:

$$m_1 - m_2 = -2.5 \log\left(\frac{F_1}{F_2}\right)$$

where
m_1 = magnitude of star 1
m_2 = magnitude of star 2
F_1 = flux from star 1
F_2 = flux from star 2

The flux received at Earth of a star which has intrinsic luminosity, L, is proportional to the inverse square of its distance:

$$F = \frac{L}{4\pi d^2}$$

where
F = flux
L = luminosity
d = distance

The magnitude system has been split into two separate systems. Apparent magnitude is the magnitude of the star as it appears in the night sky with no consideration given to how far away it is. Absolute magnitude is the magnitude of the star if it were located at a distance of ten parsecs. Apparent magnitude is easily measurable and absolute magnitude is an indication of intrinsic luminosity of the star. Thus if m_1 is the absolute magnitude, M and m_2 is the apparent magnitude, the flux ratio in the Pogson equation can be re-written as:

$$\frac{F_1}{F_2} = \frac{\dfrac{L}{4\pi(10)^2}}{\dfrac{L}{4\pi d^2}} = \frac{d^2}{10^2}$$

Thus Pogson's equation can be re-written as:

$$M - m = -2.5 \log\left(\frac{d^2}{10^2}\right)$$

This simplifies to

$$M - m = -5\left(\log d - \log 10\right)$$

which yields the distance modulus equation:

$$M - m = 5 - 5\log d$$

Providing that the absolute magnitude can be measured, perhaps by analysis of the star's spectrum, this equation can be rearranged to give an estimate f the object's distance.

its entirety was only 100,000 light years across so spiral nebulae were indeed collections of stars, similar to own Milky Way but at extreme distances. They were all galaxies in their own right.

The term nebula was dropped and the term galaxy came into use. In fact, we now know that the calibration used by Hubble for the Cepheid variable's period-luminosity relationship was wrong. Using the corrected figures, Hubble's already immense values have to be multiplied by a factor of at least two!

It was during the 1920s that the work he began during his doctorate reached its culmination. In his thesis Hubble had made notes of several important conclusions to which he would return in later life. Among those were that galaxies appeared to cluster together. Hubble also ordered the galaxies according to their shapes and placed them on a diagram (see figure 4.2). He identified five types of galaxy: elliptical, lenticular, spiral, barred spiral and irregular.

The classification of galaxies

Hubble derived seven sub-classifications for the elliptical collections of stars he observed, which depend upon how much they deviate from perfect spheres (see mathematical note 4.2). His studies had also shown that spiral galaxies formed a natural dichotomy. Although a typical spiral galaxy contains fewer stars than an elliptical, it is composed of an ellipsoidal nucleus and a surrounding disc of stars. Within that disc are spiral arms made up of very bright, high-temperature stars. In some examples, the spiral arms spin out of the nucleus directly whilst on others the arms emanate from small bars which protrude from the galaxy's nucleus. Hubble called this latter type barred-spirals to distinguish them from the true spirals. In addition, each type of spiral galaxy is sub-classified into three groups according to how large their nucleus is and how tightly wound are their spiral arms.

Lenticular galaxies contain a nucleus and a disc, similar to spiral galaxies, except that the disc does not contain any spiral arms. Irregular galaxies are just amorphous aggregates of many stars.

Mathematical note 4.2

Classification of elliptical galaxies

An elliptical galaxy will generally possess two axes of symmetry. The longer of these is known as the major axis whilst the shorter is the minor axis.

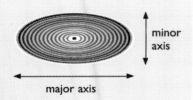

minor axis

major axis

The ellipticity of the galaxy is then easily calculated using:

$$E = 10\left(1 - \frac{b}{a}\right)$$

where
a = major axis
b = minor axis

The Hubble classification is then the nearest integer in the range 0 to 7.

Figure 4.2
The Hubble tuning fork diagram

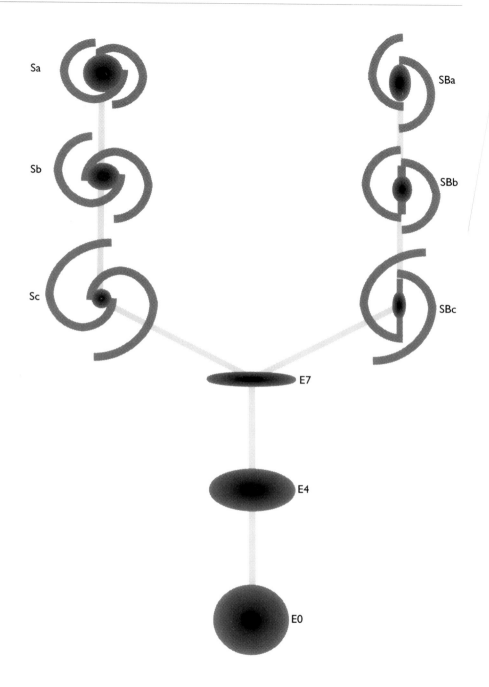

Sa

Sb

Sc

SBa

SBb

SBc

E7

E4

E0

They can also be sub-classified into those which show a vestige of spiral pattern and those which are truly irregular in shape.

Hubble felt certain that elliptical galaxies evolved into spiral galaxies and ordered his classification diagram accordingly. This double-pronged appearance has led to its being called the Hubble tuning-fork diagram and, although we now believe that Hubble was incorrect in his assumptions that ellipticals evolve into spirals, his diagram is still in wide use today as a means of classifying galaxies.

The distance-redshift relationship

As Hubble continued his work on estimating the distances to galaxies, he became aware of a trend in the data. It seemed that the higher the radial velocity of the galaxy, as indicated by its redshift, the further away the distance determination methods placed it. In 1929, he published a paper in the Proceedings of the National Academy of Sciences which showed this trend diagramatically for a sample of galaxies out to two million parsecs (see figure 4.3).

In his attempts to prove that this was, indeed, the case for the universe at large, Hubble enlisted the help of fellow Mount Wilson astronomer, Milton Humason. Humason was a skilled observer who learnt everything he knew 'on the job' because he had had no formal astronomical training. His first job at the observatory had actually been as the janitor! Humason concentrated upon recording faint galaxies' spectra and calculating their radial velocities by measurement of their redshifts. Hubble, meanwhile, spent his time estimating the distances to those galaxies for which Humason had obtained spectra. Two years after Hubble's initial paper, the two jointly published their latest results in the Astrophysical Journal. They showed that the redshift-distance relationship held true for galaxies out to and beyond 30 million parsecs.

A pivotal factor in Hubble and Humason's proof of the redshift-distance relationship was Hubble's presentation of his distance determination method. In proving that the Andromeda galaxy lay outside the boundaries of the Milky Way, Hubble had

Figure 4.3
The distance-redshift relationship

Diagram adapted from Hubble's 1929 paper in *Proceedings of the National Academy of Sciences*

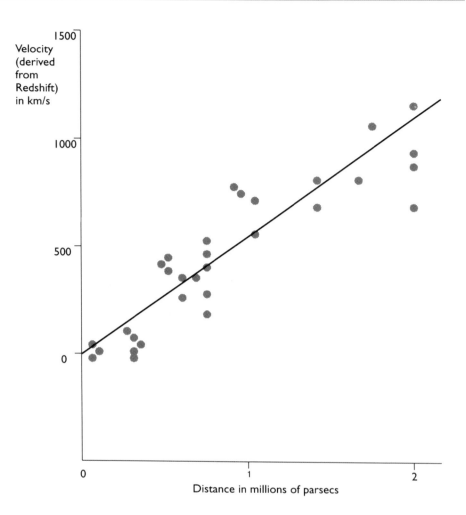

Distance in millions of parsecs

compared the different degrees of brightness possessed by Cepheid variable stars. In order to extend the distance scale to galaxies which were too far away for Cepheids to be discernible, Hubble had to change his approach a little.

Having calculated the distance to M31 based on his Cepheid variable star work he estimated the average luminosity of the brightest stars in M31. He also estimated the total luminosity of the galaxy as a whole. He then made an assumption that the brightest stars in all galaxies would be approximately the same and that all spiral galaxies would approximately be the same size and brightness. Comparing the different degrees of brightness possessed by M31 with other, more distant, galaxies, he calculated how far away they must be in order to appear that much fainter. In essence, he followed the same pattern of analysis that he did with his Cepheid variable star work but simply used a different type of object to be his distance indicator.

The Hubble constant

Having proved the existence of the redshift-distance relationship, it was immediately obvious to Hubble that, in principle, a galaxy's redshift could be measured and then converted into a distance. Furthermore, since the redshift-distance relationship was linear, i.e. it could be represented as a straight line on a graph, the factor necessary to make this conversion could be determined very easily by calculating the gradient of the line. This number has become known as the Hubble constant even though it actually changes with time (see *mathematical note 4.3*). Its value, however, is one of the most hotly debated subjects in modern cosmology.

Hubble, obviously, was the first to estimate its value. He derived a figure of five hundred and fifty kilometres per second per megaparsec. The rather peculiar dimensions of this number (often abbreviated to just km/s/mpc) indicate that a galaxy's speed of recession increases by 550 km/s for every million parsecs further out into space it is located. Thus, once the radial velocity of a galaxy is calculated from its redshift, its distance can be easily calculated by dividing the velocity by the Hubble constant.

Astronomers have gradually refined the value of the constant over the years and have now shown that it almost certainly lies between the values of 50 and 100 km/s/mpc. This means that all distances based on the redshift are uncertain by a factor of two. Periodically, lower, or even higher, values are reported.

The variability in the value of the Hubble constant is introduced,

---

Mathematical note 4.3

The Hubble parameter

In an expanding universe the force of gravity between its contents will act to slow the velocity of this expansion. Since the Hubble constant is a ratio between the expansion velocity and the distance of galaxies this, too, will become smaller with the passage of time. By visualising the Hubble constant at a specific time as the gradient of a tangent to the graph of expansion velocity versus time it can be seen from the graph that $H_f < H_0 < H_p$.

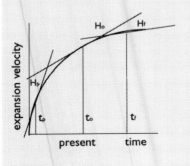

in part, because of the different ways astronomers calibrate their measurements. In order to calibrate the distance-redshift scale, some method of estimating the distances to nearby galaxies is necessary. Then these distances can be converted into the Hubble constant by using the measured redshifts. Hubble himself, as we have seen, calibrated his scale by estimating the distance using three methods: Cepheid variables, the brightest stars in the host galaxy and the brightness of the entire galaxy. Other astronomers have developed different approaches which all lead to slightly different values for the Hubble constant. There are also a number of other factors which affect the redshift and have to be removed, as will be considered later.

The very latest estimates are based upon Hubble Space Telescope measurements of Cepheid variable stars in galaxies contained within the Virgo and Fornax clusters. The analysis was spearheaded by Dr. Wendy Freedman of the Carnegie Institution of Washington. Under her guidance, HST first identified Cepheids in M100, a spiral galaxy in the Virgo cluster (see *plate 4.1*). In a carefully planned operation, the space telescope was used to take twelve one hour exposures at various times over a two month interval. From this suite of images, the regular pulsation of the Cepheid variables was identified and measured. These were then used to calculate the distance modulus to M100. The distance calculated was between 50 and 62 million light years. Combining this with its redshift gives a value of the Hubble constant of between 64 and 97 km/s/mpc.

The same analysis by the same team has also been applied to NGC 1365 in the Fornax cluster (see *plate 4.2*) which yielded a Hubble constant of between 68 and 78 km/s/mpc. This can not be taken as definitive, however, because in the interim between the Virgo and Fornax announcement, an independent team, headed by Alan Sandage, of the Carnegie Observatories, used other methods to estimate the Hubble constant and found it to be just 57 km/s/mpc.

Plate 4.1
Spiral Galaxy M100

The three inserts show a pulsating Cepheid variable star which can be used to calculate the distance to this galaxy.

(W. Freedman and NASA)

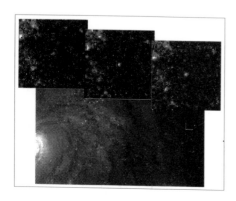

Mathematical note 4.4

The Hubble parameter

The standard non-relativistic Doppler effect was shown in mathematical note 2.1 to be:

$$z = \frac{v}{c}$$

where
v = velocity
c = speed of light

If this formula were true for all velocities, it would mean that no object could have a redshift of more than 1 because nothing can travel faster than the speed of light. We know from observation, however, that certain active galaxies can have redshifts greater than four. Thus, the formula must obviously be modified for high velocities because at speeds which are greater than ten per cent that of light, relativistic effects take place. To take this into account the special relativistic form of the Doppler effect must be used, as derived in mathematical note 2.8:

$$z = \left[\frac{\left(1 + \frac{v_r}{c}\right)}{\sqrt{\left(1 - \frac{v^2}{c^2}\right)}} \right] - 1$$

where
z = redshift
v_r = velocity in radial direction
c = speed of light
v = overall velocity

This takes the phenomenon of time dilation into account and causes the graph to have an asymptote at 1 (see figure 4.4). This equation can also be rearranged to give the separation velocity from any redshift:

$$\frac{v}{c} = \frac{(1 + z)^2 - 1}{(1 + z)^2 + 1}$$

Thus, a distant radio galaxy with a redshift of 3 is not receding from us with the impossible velocity of three times the speed of light. Instead, using the above equation, its separation velocity can be easily calculated and shown to be a factor of 0.88 times the speed of light.

special relativity to the universe at large. The predominant reason for this caution is that special relativity is only valid over volumes in which the spacetime continuum is flat. This is because a curved spacetime continuum creates a force, gravity, which must be taken into account and only general relativity can do that. Astronomers are still undecided whether or not the amount of matter contained within the universe has created an overall curvature to the spacetime continuum. They are all agreed, however, that if it is curved, it is not curved by very much. Hence, the simplification of using special relativity to analyse redshifts persists.

Another complicating factor is that at large redshifts the concept of distance becomes rather strained. This is because the separation velocities become very large and the distance to the galaxy changes significantly during the time taken for the light rays to travel to Earth. If you imagine a photon of radiation to be a runner in the one hundred metre sprint, it would be like starting the race with the finish line at one hundred metres distance but whilst you were running down the track, the finish line would be moving away from you. When you eventually caught up with the finish line, the race would end but to call it a race over one hundred metres would be meaningless.

The further away in the universe astronomers probe, the greater the separation velocities they find. By extending this line of thought it seems obvious that, somewhere in the depths of the universe, there should be regions which are so far away that they are separating from us at velocities greater than light. The very simplest form of this argument would be to divide the speed of light by the Hubble constant to see at what distance separation velocities reach this universal speed limit. Regions of the universe beyond this distance, known as the Hubble radius, are forever unknowable to us and can never be detected. Neither can the material objects in these places exert any influence on us. This introduces us to the idea of a universal horizon. We have already encountered one form of redshift horizon in the previous chapter: that of the event horizon around a black hole. This time we are talking about a horizon which surrounds us at a vast distance. Whilst it is possible to derive precise equations which give the horizons (see mathematical note 4.5), it is as yet impossible to calculate them

Mathematical note 4.5

The Hubble radius and horizons

The Hubble radius is defined to as:

$$r_H = \frac{c}{H_0}$$

where
r_H = Hubble radius
H_0 = Hubble constant
c = speed of light

This crude, linear estimate is far too simplistic in real life and much more rigorous equations can be derived which show that the distance, σ_{oh}, of the furthest visible object or event in the universe is:

$$\sigma_{oh} \propto c \int_0^{\tau} \frac{dt}{R(t)}$$

where
σ_{oh} = distance of further observable object/event
τ = age of the universe
c = speed of light
$R(t)$ = scale factor at time, t

The precise equation depends upon the overall curvature of the spacetime continuum.

with any precision because the overall curvature of the universe is still unknown.

Keeping the fact that distances are so uncertain in mind, it can be seen that to talk about a galaxy existing at a certain distance requires an astronomer to then qualify his statement by explaining what value of the Hubble constant he is using. Perhaps he would even need to define which method of distance determination he preferred. Anyone wishing to use his data would necessarily have to convert all distances using their own preferred value of the Hubble constant. To avoid this problem, astronomers have adopted the convention of referencing galaxies by their redshifts rather than by giving distances.

The expanding universe

Hubble's and Humason's results were extremely important because they seemed to imply that the whole of the universe was rushing away from us as if, to coin a popular cosmological phase, our Galaxy were possessed of some appalling cosmic body odour problem! Furthermore, the fact that everything was moving away from the Galaxy seemed to imply that the Milky Way was situated at the centre of the universe. This rang alarm bells in many astronomers' heads because, throughout the whole history of astronomy, many of the major advances in our understanding had been made by divesting our position in the universe of a level of importance. Now to place us centre stage after all, smacked of a backwards step. The resolution to the problems thrown up by Hubble and Humason's results were tied up in Einstein's general theory of relativity. In fact, some theorists were already exploring the ideas.

One year after its formal presentation, Einstein set about applying his general theory to the universe as a whole. In 1917, the conventional wisdom of that time stated that the universe was static. It is a little peculiar that this should have been the case because Newton himself had realised that his laws of gravitation must have some effect on the universe as a whole. In particular, he noted that it was strange that the universe had not collapsed because of the force of gravity between the objects within it. He chose not to pursue this line of reasoning, however, and the subject matter was all but forgotten.

Despite many attempts, Einstein found it impossible to produce a solution to his equations which gave him a static, unchanging cosmos. Even if he assumed the universe was infinite in size, his equations still predicted that space itself should be in motion. Either expansion or contraction were predicted to be possible but not a stationary universe.

Understandably disappointed at this perceived failure of his theory, Einstein introduced a new term into the equations. He called it the cosmological constant and used it to balance the attractive force of gravity and hold the universe static. The introduction of the cosmological constant was a neat mathematical trick but a trick nonetheless. It was an ad-hoc addition to the theory which forced it to fit the facts. As it turned out, the so-called facts were nothing more than preconceived ideas about the state in which the universe existed!

Later investigators of relativity found that, even with a cosmological constant, the universe could still be expanding. Three of the most important contributions to this study were made by Dutch astronomer Willem de Sitter who showed that a fundamental property of spacetime was that it expanded, Russian mathematician Aleksandr Friedmann, who defined that the spacetime continuum would either expand forever or would eventually re-collapse and the Belgian priest and cosmologist Georges Lemaître, who speculated about the origin of the universe based on the fact that if it is expanding, it must once have been very small. His work allowed him to predict the expanding universe two years before Hubble announced the recession of the galaxies. When Einstein learned of Hubble's work which showed the expanding nature of the universe, he immediately stated that the introduction of the cosmological constant was the biggest blunder of his life!

The work of these cosmologists suggests that it is the spacetime continuum which is expanding, dragging the galaxies along with it. In effect, this means that the galaxies are actually stationary within the spacetime continuum rather than rushing through it *(see figure 4.5)*. Imagine that the universe is covered by a regularly spaced grid with each galaxy sitting at a co-ordinate. Then, as the universe expands, so the co-ordinate grid stretches as well. Although, the distance between the galaxies has increased, the galaxies all still possess the same co-ordinates.

As spacetime expands it stretches the rays of electromagnetic radiation passing through it, causing a redshift. In correct scientific terminology the universe is said to be in a state of uniform expansion. This means that the expansion is taking place everywhere at a constant rate. An earth-bound example of uniform expansion is bread rising in an oven *(see figure 4.6)*. The most basic property of a uniformly expanding medium is that the greater the distance between two objects, the faster they will be separating *(see figure 4.7)*.

Using this notion of uniform expansion, it suddenly becomes easy to explain why the Milky Way appears to be the centre of the universe. Imagine a balloon with polka-dots scattered over its surface. The balloon represents the spacetime continuum, whilst the dots represent the galaxies. After you have blown the balloon up a little, select one of the dots and put your finger on it. Now blow the balloon up even further and try to imagine what it looks like to your finger. From its vantage point it looks as if every other polka dot is moving away from it.

Figure 4.5
The expansion of spacetime

In these diagrams the spacetime continuum is represented by the grid upon which the galaxies are positioned. It is incorrect to think that the expanding universe is produced by the galaxies moving through the spacetime continuum as in the top diagram. Instead, spacetime is expanding and 'dragging' the galaxies along with it as shown in the bottom diagram. To fully understand the difference notice the way that the galaxies change their position in the grid in the top diagram but remain in fixed positions, even though the grid has got larger, in the bottom diagram.

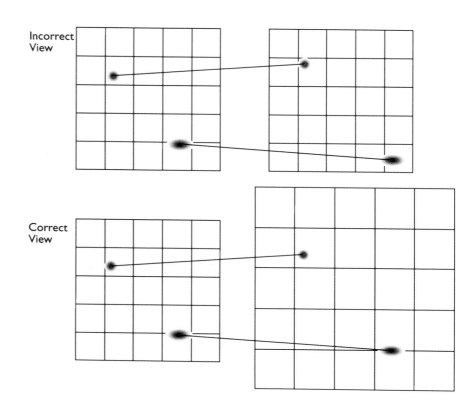

Incorrect View

Correct View

Figure 4.6
The currant bun universe

This diagram represents dough with currants scattered throughout its volume. After it has been baked in an oven, the mixture has risen to twice its original size. The currants have remained the same size but the dough in between has doubled in volume. Thus, the currants are now twice as far away from each other as they were originally. All the currants have remained static relative to the dough mixture and from any currant's perspective, it appears to be the other currants which have moved. This is a good analogy to the real universe, the currants are galaxies whilst the dough mixture is the spacetime continuum!

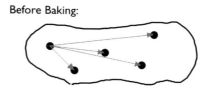

Before Baking:

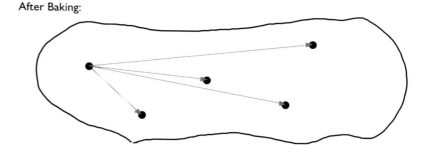

After Baking:

Figure 4.7
Uniform expansion

In this example, the top ruler has taken two minutes to expand and become the bottom ruler. Thus, from the given velocities it can be seen that the speed of expansion is proportional to the distance from the centre of the expansion, in this case the zero mark. This proportionality, i.e. if you double the distance you double its velocity, is the basic property of a uniformly expanding medium such as the spacetime continuum.

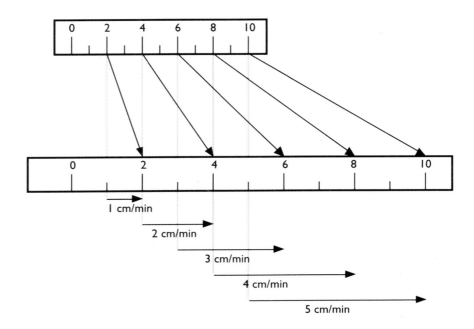

Select another dot and perform the experiment again and continue until you convince yourself that no matter which dot you choose, it always appears to be the centre of the expansion. It is actually an 'optical illusion' brought about by the fact that every galaxy is stationary within the spacetime continuum.

As with everything in nature, things are not quite that simple and the above explanation is an over-simplification necessary in order to convey the principal ideas. In reality, on the scales studied by astronomers so far, the universe itself is not a laboratory example of uniform expansion. The exact rate of the universe's expansion changes throughout space depending upon the density of matter in the expanding region. Mass creates gravity, an attractive force, which tries to pull things closer together. It therefore resists the expansion of the universe which tries to force everything apart. To return to the balloon analogy it is as if the balloon's skin is slightly thicker in certain places and not so easy to stretch.

In fact, in certain regions of the universe, the cosmic expansion can be overcome completely by associations of galaxies which are bound together by gravity. These are known as groups or clusters depending upon how many galaxies are in the association. The Milky Way is gravitationally bound to the Andromeda galaxy and together their mutual force of gravity is enough to overcome the expansion of the spacetime continuum. At the current time, our galaxy and Andromeda are actually moving towards one another. This explains why Slipher detected a blue shifted spectrum when he studied M31. These two spiral galaxies are thought to dominate the so-called Local Group which is a collection of some thirty or so galaxies, of different types, mostly irregulars and dwarf ellipticals. All of them are in orbit around the group's centre of mass. The Local Group itself moves in response to the distribution of matter in the universe around it. Thus, if we are trying to determine exactly the rate of universal expansion we have to first understand how fast and in what direction the Local Group is moving. Luckily enough, the universe is bathed in the cosmic microwave background radiation which was released when the universe was very young. This radiation is being perpetually stretched as the universe expands and so gives astronomers a very handy universal frame of reference against which to measure our velocity. Our motion imparts a small Doppler shift upon it so that in one direction it is blueshifted whilst in the other it is redshifted. The blueshift indicates the direction which we are moving towards. When the Doppler shift measurement is used to calculate velocities, the Sun appears to be moving in the direction of the constellation of Leo with a velocity of 370 km/s at the present time.

This, unfortunately is not the end of the problem. This figure is not just composed of the Local Group's motion, it also contains elements from our galaxy's motion within the Local Group and the motion of the Sun around our galaxy! When these other factors are estimated and removed, the motion of the Local Group is all that remains. Current estimates show that the Local Group is moving in the direction of the constellations Hydra and Centaurus with a velocity of 600km/s.

When we measure the cosmological redshift of distant galaxies it is very important to take into account the motion of the Local Group because, in most cases, it will impart an additional Doppler shift which will cause the redshift to be either over or underestimated (see figure 4.8). In recent years, astronomers have also begun to suspect that another systematic Doppler shift is being imparted to a very large number of galaxies which are close enough for us to study. It appears as if many clusters of galaxies, our own Local Group included, are being pulled towards a incredibly massive conglomeration of matter. It has become known as the Great Attractor but its composition remains a mystery. It is suspected to be a titanically huge supercluster of galaxies but it is very difficult to observe because our line of sight to it passes straight through the disc of the Milky Way where stars and dust obscure an astronomer's view. In fact, it is so difficult to observe through the galactic disc, that the swathe cut by the Milky Way across the night sky is known to astronomers rather dramatically as the Zone of Avoidance. This type of large scale, systematic motion has been termed streaming.

Despite this, astronomers still believe that on its largest scales, the universe can be thought of as homogeneous. This means that the local density enhancements can be 'averaged out' with corresponding areas of depleted density and the distribution of matter can be thought of as being smoothly uniform (see figure 4.9). Thus on its very largest scales, astronomers still hope that the universe behaves in a uniformly expanding way. As yet, even the largest scales studied are too small to show this behaviour conclusively.

Figure 4.8
Redshift distortions caused by motion of the local group

This schematic representation indicates the clusters' masses by their areas and the directions of the motion by arrows: the longer the arrows, the larger the velocites. Thus when we are measuring redshifts from the Local Group out to cluster A, we will underestimate the true reading because we are being pulled towards it by its gravity.

Cluster B also exerts a gravitational pull upon us and that is why the Local Group's motion is not directly towards the centre of cluster A. Cluster B, on the other hand, will have its redshifts overestimated because the Local Group is moving away from it.

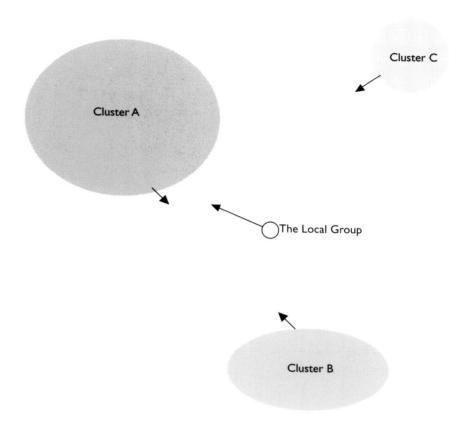

Without a shadow of doubt, the complications brought about by the local distribution of matter make it very difficult to calibrate the expansion of the universe by quantifying the Hubble constant. It is hoped that, if we can fully elucidate the distribution of matter in the universe around us, a more precise understanding of the cosmic expansion will be forthcoming. The best way to survey the surrounding galaxies and build maps is by measuring their redshifts.

Mapping the universe

Anyone who has looked through a telescope knows that it is very difficult to gauge distances in the universe around us. The simple act of looking at a piece of sky gives us no clue about the

Figure 4.9
The cosmic substratum

The universe is assumed to be homogeneous at some large scale. When dealing in those scales, the matter contained in the universe can be thought of as being spread uniformly throughout space. It is then known as the cosmic substratum. The shaded ellipses in the left hand column represent the density of galaxies in different regions of space. The darker the ellipse, the denser the galaxies. The overlying circles are the regions over which the density is averaged. Only when the circles in the right hand column are the same colour can the universe be thought of as homogeneous.

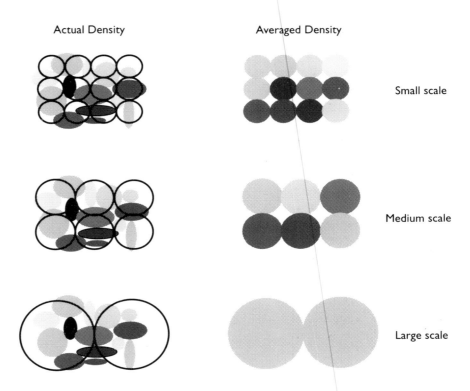

Actual Density   Averaged Density

Small scale

Medium scale

Large scale

distances to the various objects we are seeing. Everything appears to us as if it has projected onto a giant crystal ball surrounding the Earth. This has been termed the celestial sphere although smaller areas of it are often referred to as the sky plane. Celestial objects all appear to be sitting on the celestial sphere and pictures of the night sky can often show stars, galaxies and quasars, all of which actually exist at vastly different distances (see figure 4.10). The classic examples of this are the constellations themselves. They are simply chance alignments of stars which, in reality, are nowhere near to one another. In many cases, dim stars appear bright because they are close to us whilst bright stars appear dim because they are far away.

In order to make some sense of what we are seeing, a way has to be found of calculating the distances to celestial objects. This

Figure 4.10
Projection onto the celestial sphere

The rectangular box shows what the observer can see with the plane of this page representing the sky plane. By looking at how the objects have been projected onto the celestial sphere, the 'double star' visible in the top right hand corner can be seen to be an optical illusion. Similarly, the observer cannot tell just by looking that the two galaxies in the field of view are at a vastly increased distances than the stars.

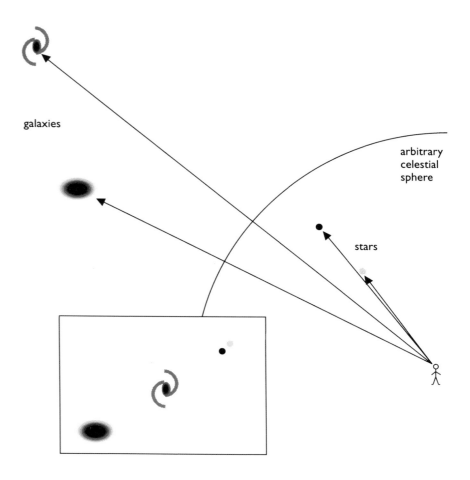

then gives us a truer picture of the universe in which we live. As the previous sections have explained, the problem is that, beyond the local universe, distances are very uncertain and, if you look far enough, even the concept becomes unclear. So the redshift is used instead because astronomers know that this is an absolute property which is related in some way to an object's distance even if they do not yet know the precise relationship. Thus the way the universe is mapped is by taking an object's spectrum and measuring its redshift. It is the redshift which is then plotted instead of the distance. Maps made in this fashion are known as maps in redshift space and have produced a variety of fascinating results.

The fact that galaxies cluster together has been known for a fact since Hubble conducted statistical analysis of the distribution of galaxies on the Mount Palomar photographs. Clyde Tombaugh, the discoverer of Pluto, noticed that these clusters also gather together into superclusters. The real extent of the clustering, however, has only become apparent with the advent of redshift mapping. It would appear that the universe is clustered on the largest scales so far studied.

In redshift space, clusters of galaxies appear as elongated structures like vast fingers pointing towards the Earth *(see figure 4.11)*. Some astronomers have whimsically termed these structures, fingers of God. In actual space, however, the clusters are spherical. The reason for this is that each galaxy has a Doppler shift caused by its motion around the cluster's centre of gravity, superimposed upon the cosmological redshift *(see figure 4.12)*. For the relatively nearby clusters, these Doppler shifts and the uncertainties in their measurement, can have a very noticeable effect when combined with the cosmological redshift of the cluster.

The superclusters of galaxies form sheets and filaments which surround vast regions of space, known as voids, which are almost devoid of galaxies. Although it is very difficult to do so accurately, it is almost impossible not to try and convert some of the redshift measurements into distances. When this is performed the largest voids appear to have diameters of approximately 150 million light years. Some galaxy agglomerations are so massive that they have achieved fame in the astronomical community and have been nicknamed. For example, the Great Wall is the largest known structure in the universe. It is estimated to contain thousands upon thousands of galaxies and to extend over an area of 500 million by 200 million light years. In terms of thickness, however, it is only about 15 million light years.

Figure 4.11
Clusters of galaxies in redshift space

This triangular wedge of redshift space has its apex centred on the Local Group. The Coma 1 Cloud is actually part of the Local Supercluster. Beyond that lies the vast Coma cluster, elongated on this diagram because of redshift distortions brought about by the individual velocities of the component galaxies. The Coma supercluster has been marked by the light grey swath which cuts across the diagram encompassing several other groups as well as the Coma cluster.

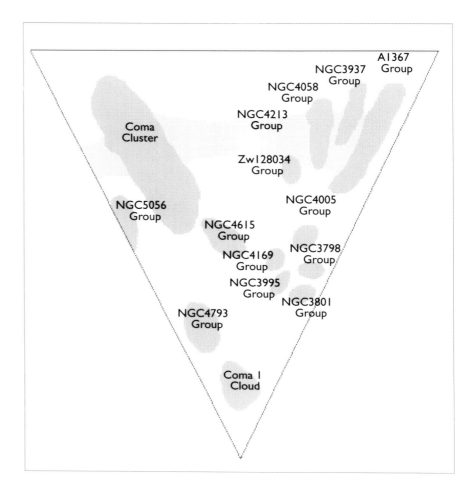

When using clusters of galaxies to estimate the value of the Hubble constant it is far better to use compact clusters rather than large clusters. For example, the Virgo cluster of galaxies is roughly at the same distance from our Local Group as the Fornax cluster of galaxies. Up until now, the Virgo cluster has usually been chosen for study because it contains a large number of galaxies, 1,170 in fact. The drawback to this is that the volume of the cluster is so great that the distance to any one galaxy, ignoring any redshift distortion caused by the galaxy's own motion around the cluster's centre of mass, is not a reliable estimate of the cluster's average distance (see figure 4.13). It is the average distance of the cluster which is needed for an

Figure 4.12
Redshift distortions in clusters

When the redshifts of galaxies are measured, superimposed upon the cosmological redshift is a Doppler shift caused by the galaxies' individual motions around the cluster's centre of mass. The circular diagram represents the actual cluster in space with the arrows representing the galaxies' velocities. The elliptical shape is the way the cluster appears in redshift space because it is impossible to distinguish a Doppler shift from a cosmological redshift. If the galaxy is moving towards us, its redshift is underestimated whilst if it is moving away, its redshift becomes exaggerated.

Only by taking a large number of galaxies' redshifts and averaging them out can the redshift of the cluster be precisely known.

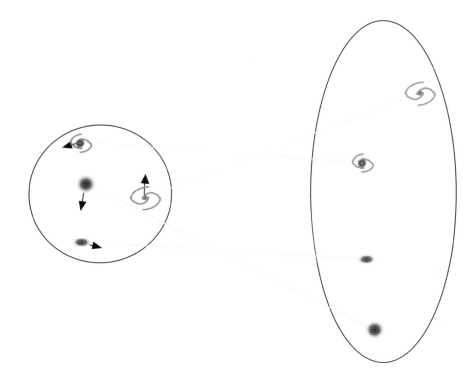

accurate estimate of the Hubble constant. With a smaller cluster, such as Fornax which contains only 235 galaxies, the volume of space occupied by the cluster is much less and so the Hubble constant estimates are far better.

Interpreting the cosmological redshift

On its largest scales the universe is the realm of general relativity. We have considered that the spacetime continuum can be thought of as a grid of co-ordinates, expanding with the universe and that the clusters of galaxies do not change their position relative to one another, only their separations. In order to keep track of this expansion we must define a parameter known as the cosmic scale factor. This is an arbitrary measure of the size of the universe. In a static cosmos, the scale factor

Figure 4.13

The average distance to clusters

When determining the distances to clusters of galaxies, small, compact, clusters are preferable to large ones. This is because in a small cluster, even the most outlying galaxy is still a reasonable representation of the cluster's distance as a whole. In a large cluster of galaxies, the distance can be affected by millions of light years if the chosen galaxy happens not to be located near the cluster's centre.

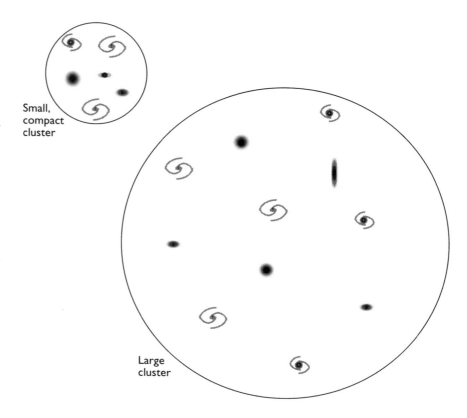

Small, compact cluster

Large cluster

would be a single constant figure. In our expanding one, however, the scale factor perpetually changes with time. The precise way in which it changes is unknown although a number of possibilities exist.

In this formulation, the amount by which the wavelength of radiation has been stretched is directly linked to the amount by which the universe has expanded. In fact, when the value of an object's cosmological redshift is incremented by one, the result is a quantity known as the spectral ratio. This figure represents the factor by which the universe has increased in size since the object under scrutiny released its radiation. For example, if a galaxy were discovered to possess a redshift of 1.0, that means the light currently being detected from it was emitted when the universe was half its present size (see *mathematical note 4.6*).

Mathematical note 4.6

The scale factor

The scale factor is an arbitrary measure of how big the universe is at any given time. As the universe expands so the wavelength of radiation it contains is also stretched. Thus it follows that, in some way, the wavelength of emitted radiation and the scale factor must be linearly proportional to one another:

$$R \propto \lambda$$

where
R = scale factor
λ = wavelength

If R_0 is the present day value of scale factor and λ_0 is the wavelength of radiation which was originally released with a wavelength of λ when the universe possessed a scale factor of $R(t)$, then:

$$\frac{R(t)}{R_0} = \frac{\lambda}{\lambda_0}$$

where
$R(t)$ = scale factor at time, t
R_0 = present day scale factor
λ_0 = emission wavelength
λ = present day wavelength

Remembering from mathematical note 2.1 that:

$$\frac{\lambda}{\lambda_0} = 1 + z$$

combining the two equations gives:

$$\frac{R(t)}{R_0} = 1 + z$$

where
$1 + z$ = spectral ratio

The spectral ratio is the factor by which the universe has expanded since time, t, to the present day.

Furthermore, a redshift of 1.0 implies a very large separation velocity between the galaxy in question and the Milky Way. In fact, according to the simplest general relativistic models of the universe, a redshift of 1.0 would imply that the spacetime continuum between us and the distant galaxy was expanding at the speed of light. A galaxy with a redshift of 2.0 would imply that the space in between us and it was expanding at twice the speed of light! At first, these fantastic statements appear to contravene the law that nothing can travel through the universe faster than light. In fact, nothing is travelling through the universe

Figure 4.14
Look-back time

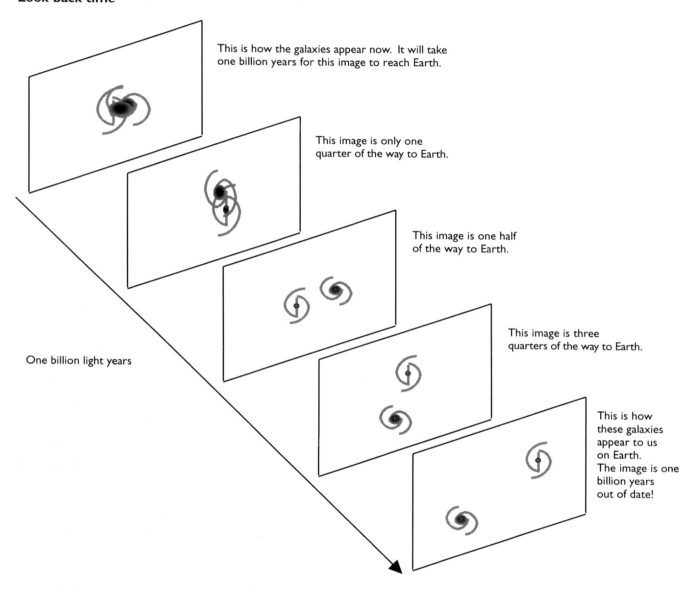

This is how the galaxies appear now. It will take one billion years for this image to reach Earth.

This image is only one quarter of the way to Earth.

This image is one half of the way to Earth.

This image is three quarters of the way to Earth.

This is how these galaxies appear to us on Earth. The image is one billion years out of date!

One billion light years

Look-back time is a consequence of the finite speed of light. Light travels at approximately 300,000,000 metres per second but even so, the distances in space are so vast that light can take millions or even billions of years to reach us. In these enormous spans of time, the original objects will have changed and evolved. Astronomers have devised the concept of a light year which is the distance light travels in a year. If an object is one billion light years away, we see it as it was one billion years ago. There is no way for us to know what it looks like today, faster than the speed of light because it is the universe which is expanding whilst the galaxies remain stationary within it (apart from their peculiar motions within their clusters). The spacetime continuum, itself, can expand with whatever velocity it likes and no physical laws are violated.

The evolving universe

Galaxies are so far away that even light, the fastest thing in the universe, takes a great deal of time to cross the distances involved. The consequence of this is that if the light from a certain celestial object has taken one million years to reach us, then the image we see is how the object appeared one million years prior to the observation (see figure 4.14). To give some idea of the scales with which we are dealing: the Andromeda galaxy is the nearest major spiral galaxy to the Milky Way. It is located 2.2 million light years away which means that its light takes 2.2 million years to get to us. Thus, we see the Andromeda galaxy as it appeared 2.2 million years ago.

In effect, when astronomers look out into space they are seeing what the universe used to look like. The further away they look, the earlier in its evolution they are seeing. This phenomenon is known as look-back time. Again, because distance is subjective when talking about the far reaches of the universe, redshifts are usually used. By taking a 'census' of astronomical objects and correlating them against redshift, it becomes immediately obvious that the contents of the universe have evolved and changed with time (see figure 4.15).

Marking the limits of the observable universe is the cosmic microwave background radiation. This possesses a redshift of approximately 1,000 and means that the universe was once dominated by very energetic radiation. Nowadays, because the universe has expanded its volume so much, by about 1,001 times, the radiation has been redshifted down into the microwave region of the electromagnetic spectrum.

Figure 4.15
The Universe in redshift space

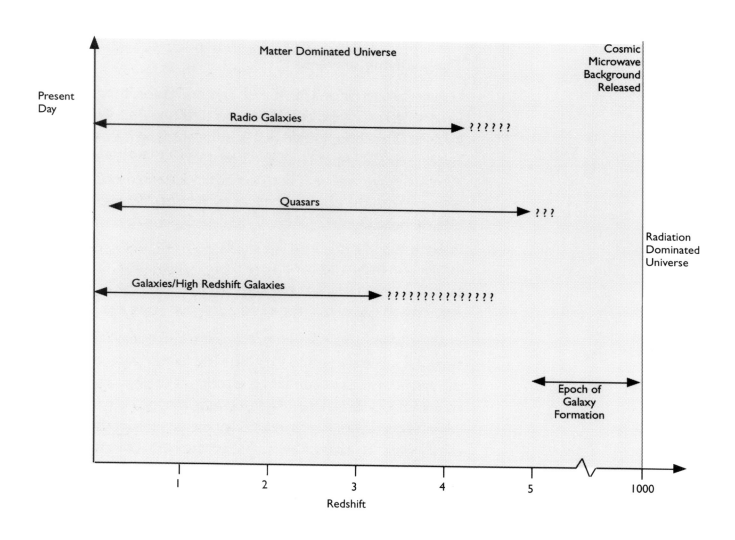

The release of the cosmic microwave background radiation represents the limit of the observable universe. Before this event, which took place approximately 300,000 years after the universe was created, matter was uniformly spread throughout the cosmos and constantly buffeted by radiation. Hence, it is known as a radiation dominated universe. The release of the background radiation corresponded to the condensation of matter into neutral atoms which then grouped together and formed galaxies. The universe is now known as a matter dominated universe.

The cosmic microwave background radiation blocks our view of the first 300,000 years of cosmic history. At the point in time when this radiation was released, the universe had just become full of electrically neutral atoms. Before this, the cosmos had been filled with a mass of disorderly electrons and atomic nuclei which were so energetic that they kept hindering light's passage by colliding with the photons of electromagnetic radiation. As the universe expanded, however, these various particles slowly lost energy and began to combine into atoms. This event is known as the decoupling of matter and energy because, suddenly, the photons were no longer in a constant state of collision with the particles. The background radiation was released at this point and has been travelling through space to reach us ever since.

Superimposed on this background radiation are the 'ripples' mentioned in Chapter Three. These are caused by the gravitational redshift and can be used to trace the density of matter in the early universe. They give a clue as to how the disorderly 'soup' of particles became the galaxies of today.

Following the microwave back ground radiation, there occurs a region in redshift space which we will term the astronomical wilderness. It extends from a redshift of just below 1,000 to a redshift of approximately 5. Although this appears to be a vast range of redshifts, it is important to remember that the redshift is not a linear scale and that, in terms of time, this range only covers a period of one billion years. The astronomical wilderness is where galaxies are believed to have begun forming. As soon as matter and energy decoupled, releasing the microwave background radiation, vast clouds of gas, which show up as dense ripples in the background radiation, began to collapse under the weight of their own gravity. The universe became dominated by matter as it clumped together, forming superclusters, then clusters and then galaxies.

Until individual galaxies began to form, the nascent superclusters and clusters were nothing more than vast clouds of gas and dust. This makes them totally impossible to detect with today's technology, because they are so far away and contain no stars to illuminate them.

In an effort to determine exactly where in redshift space the majority of galaxies formed, astronomers using the Hubble Space Telescope have made a systematic study of very distant clusters. To date, the furthest, confirmed, galaxy cluster lies at a redshift of 3.33, placing it near the edge of the astronomical wilderness. These images give astronomers their first views of how galaxies have evolved into the spirals and ellipticals we see around us today.

The cluster at $z=3.33$ is in the constellation of Sculptor and clearly shows galaxies which are in the process of forming. Although spiral galaxies are not present in the cluster, galactic objects which appeared as if they might become spirals have been identified. Elliptical galaxies, however, were present and looked remarkably similar to those in the universe today. In a closer cluster, located at a redshift of 1.2, in the constellation Serpens, astronomers again found the same pattern: fully evolved elliptical galaxies but nascent spirals. The third cluster looked at in this survey was cluster CL 0939+4713. In Ursa major, it has a redshift of 0.4. In this cluster, ellipticals and almost normal-looking spirals were easily identified. This raises the interesting point that, whilst spiral galaxies appear to develop over a long period of time, ellipticals seem to be born very quickly and then change little over the course of cosmic history (see *plate 4.3*). Two other important features became obvious from this survey work as well. Firstly, the proportion of spiral to elliptical galaxies in today's universe is lower than in the past. Secondly, a large number of galaxy 'fragments' were noticed in the surveyed clusters.

Plate 4.3
Galaxy evolution

This sequence shows a spiral and an elliptical galaxy from the present universe (left) and a sequence of evolving galaxies from when the universe was only 9, 5 and 2 billion years old. Ellipticals form quickly but spirals take a long time to evolve.

(A Dressler, M Dickinson, D Machetto, M Giavalisco and NASA)

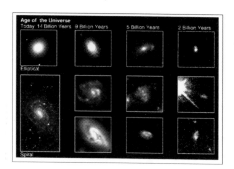

Doing more work on distant galaxies, using the Hubble Space Telescope, Rogier Windhorst of Arizona State University discovered eighteen 'sub-galactic clumps', crammed into a volume of space approximately two million light years in diameter (see plate 4.4). This galaxy building site exists at a redshift of 2.39. They have been termed sub-galactic clumps because each of the objects is only a few thousand light years in diameter, making them much smaller than a galaxy in today's universe. The central bulge of the Milky Way is about 8,000 light years across (while the spiral arms measure about 100,000 light years in diameter). Being much smaller than today's galaxies implies that, somehow, these galaxy building blocks have grown. There are two possibilities for this, the first is that they have grown through the gradual accumulation of matter, the second is that they have grown through collisions and mergers.

This intriguing question is one which can be addressed using the Hubble Deep Field (see plate 4.5). This amazing image has been termed the Hubble Space Telescope's legacy to observational cosmology and it was captured by the space telescope during December 1995. Three hundred and forty two separate images of the same region of sky were taken over a consecutive ten day period (between 18th and 28th December) and combined together to give the deepest ever image of the universe. The term 'deep' is used to indicate that the image captures objects which are very far away in space. These are usually faint because of their distance and so require long exposure times. In the case of the Hubble Deep Field, images at four wavelengths were taken so that when they were combined a coloured image would result. The four wavelengths (or photometric bands as they are called) chosen were U band; ultraviolet (300 nanometres), B band; blue (450 nanometres), V band; yellow-red (606 nanometres) and I band; infrared (814 nanometres). This choice of wavelengths was also essential for the constructive use of the data rather than just to produce a nice image.

Plate 4.4
Galaxy building blocks

Each one of these objects is about one third the size of the centre of the Milky Way. They seem destined to merge and form larger galaxies.

(R Windhorst, S. Pascarelle and NASA)

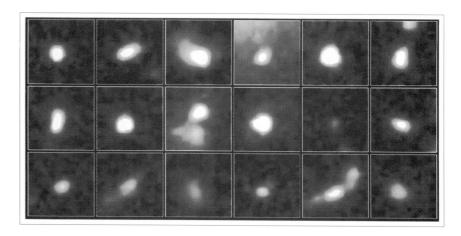

Plate 4.5
Hubble Deep Field

The deepest ever view of the universe. On this image are several thousand galaxies which span the length of cosmic history.

(R. Williams, the Hubble Deep Field Team and NASA)

Analysis of the Hubble Deep Field (HDF) is beginning to give some remarkable details about how galaxies form. Before this can be done, however, astronomers must first add depth to the HDF. In other words they must begin to work out the redshift to the many objects found in the image. Over 3000 galaxies have been identified on the HDF and each must now have its redshift calculated or measured. Although the giant 10 metre Keck telescope is spearheading an observational programme which has so far captured several hundred redshifts, to observe each galaxy spectroscopically would take an eternity and be almost impossible, because some of the objects on the HDF are too faint to be seen from the ground. Instead, a way of estimating the redshift from the different colours of the galaxies has been used. This provides something known as a photometric redshift.

The technique has been applied to the HDF by Kenneth Lanzetta of SUNY at Stony Brook, among many others. The basic method Lanzetta and his co-workers applied was to look at the colour of nearby galaxies. These were then 'redshifted' by a computer to give templates of what a galaxy should look like at various different redshifts. These templates could then be matched to a real object on the HDF in order to estimate its redshift. Using this system, incredibly distant galaxies were discovered, a handful of which appear to be further away than the most distant quasars *(see plate 4.6)*.

In general, the redshifts of the objects on the HDF are so large that what appears to be visible light was emitted as ultraviolet radiation. This means that most of the objects on the HDF are vigorously forming stars, because ultraviolet radiation is given out predominantly by high temperature, short-lived stars, closely associated with regions of ongoing star formation. This, in itself, is good evidence that most of the HDF objects are galaxies in the process of formation.

Plate 4.6
Distant galaxy in the HDF

This galaxy is thought to be so far away that it existed when the universe was only five per cent of its present age (about 1 billion years after the big bang).

(K Lanzetta, A Yahil and NASA)

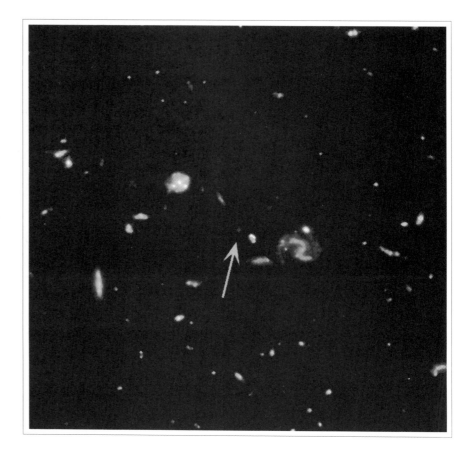

The vast majority of objects on the HDF are unlike galaxies in today's universe, i.e. they are neither spiral nor elliptical. Most appear to cluster at around redshift 2 and tail off dramatically. At redshifts greater than five, virtually nothing is visible. In an attempt to understand how galaxies form, astronomers from the University of Durham, led by Carlos Frenk, have run simulations which follow large collapsing clouds of gas through to the ignition of stars and hence the formation of a galaxy. They discovered that their young galaxies looked like many of the objects on the Hubble Deep Field. The objects would then continue to evolve but the factor which determined whether each would end up as a spiral or as an elliptical galaxy was the number of collisions it experienced during its lifetime.

Left to its own devices, a collapsing cloud of gas will generate stars quiescently and eventually become a spiral galaxy. If evolving galaxies collide and merge, this can ignite enormous bouts of star formation, known as starbursts, which quickly use up the raw, gaseous material in galaxies, throw stellar orbits into confusion and destroy the beautiful spiral patterns. The resulting, random orientation of the stars' orbits results in elliptical galaxies. This fits in with the view that spiral galaxies appear to change greatly with time, whilst ellipticals do not. As soon as the mergers happen, ellipticals cease to evolve because they rapidly run out of star forming material. It also explains why more distant clusters have a greater proportion of spiral galaxies than ellipticals, as their spiral galaxies have had less time to collide and become ellipticals.

Returning to the edges of the astronomical wilderness we can begin to explore the question of where the active galaxies fit into all of this. At around about redshift 5 the first quasars are observed. These objects are still something of a mystery in modern astronomy. Their spectra indicate that they are broad line active galaxies as described in Chapter Two but they are far more luminous than the other active galaxies. Because they are so bright they have long been the farthest detectable objects in the universe. The density of quasars in space steadily increases until it reaches a peak between redshifts 2 and 3, much like the galaxies in the Hubble Deep Field. The density then tails off until the closest known quasar which is at a redshift of less than 0.1. The reason for the peak in the density of quasars is unknown at the present time, having only recently been shown to be a real feature of the data rather than a statistical oddity. In light of some new Hubble Space Telescope results, however, it almost certainly has to do with the fact that the rate of mergers between galaxies was greater in the early universe (see *plate 4.7*). Even so, some quasars have been found in totally undisturbed spiral and elliptical host galaxies and it is interesting

Plate 4.7

Quasar host galaxies

The incredibly bright quasar cores are shown in these images to reside in both normal and merging galaxies.

(J Bachall, M Disney and NASA)

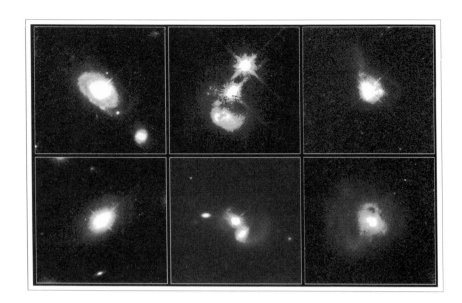

to speculate about whether all galaxies underwent a period of quasar activity. The HDF would seem to indicate that, in fact, they did not. Only two quasars in the whole field of view have been detected, out of the 3,000 galaxies on that image, by workers (at Jodrell Bank, England and the Very Large Array, New Mexico). Another type of active galaxy, known as a radio galaxy, has examples which stretch throughout space to a redshift of 4.25.

In contrast, just about every 'normal' galaxy ever observed lies between redshifts of 0.1 and 0. Those galaxies with redshifts greater than this are considered so far away that they are distinguished with the name high redshift galaxies.

The age and fate of the universe

It is important to stress that the value of the Hubble constant, calculated from the study of cosmological redshifts, is the most important figure in cosmology. It leads us to a quantity which is known as the Hubble time (see *mathematical note 4.7*). The Hubble time is the time it would take for a galaxy to double its distance from us. In other words, it is the time in which it would

---

Mathematical note 4.7

The Hubble time

The Hubble constant can be used to define the Hubble time:

$$\tau_0 = \frac{1}{H_0}$$

where
τ_0 = *Hubble time*
H_0 = *Hubble constant*

The Hubble time is a very rudimentary estimate for the age of the universe.

If H_0 is 50 km/sec/Mpc,
$\tau_0 = 2 \times 10^{10}$ years.
If H_0 is 100 km/sec/Mpc,
$\tau_0 = 1 \times 10^{10}$ years.

take a galaxy to traverse its present distance, thus doubling it. This definition implies that the Hubble time is also the time taken by the galaxy to reach its present position measured from us.

As a brief aside before continuing, this chapter has assumed that the universe began in a big bang . Indeed, the expanding universe is a strong piece of evidence in favour of that hypothesis. At the moment of the big bang, theory holds that all the matter in the universe was located at the same point in space. Thus, if we return to our definition of the Hubble time which states that it is the time taken for the galaxy to reach its present distance, it can be seen that the Hubble time actually represents the age of the universe. Depending upon the choice of Hubble constant (either 100 km/s/mpc or 50 km/s/mpc), the Hubble time becomes ten thousand million years or twenty thousand million years.

These estimates, based upon the Hubble law, are kinematic in origin and so they are overestimates because they do not take into account the force of gravity created by the matter content of the universe, which actually reduces the age of the universe to approximately two thirds of the Hubble time (see figure 4.16).

Obviously, the precise amount by which the cosmic expansion is decelerated depends upon how much matter is present in the universe. In principle, it should be possible to measure the amount of matter in the universe because it will have imparted a slight curvature to the spacetime continuum.

There are two possibilities for this curvature. If there is enough matter in the universe to create a strong enough gravitational field, eventually the expansion will be overcome and the universe will collapse. If this is the case, the universe is said to be closed and the curvature is said to be negative. If a two dimensional sheet of paper were given negative curvature it would become a three dimensional hemisphere. A positively

Figure 4.16
The age and fate of the universe

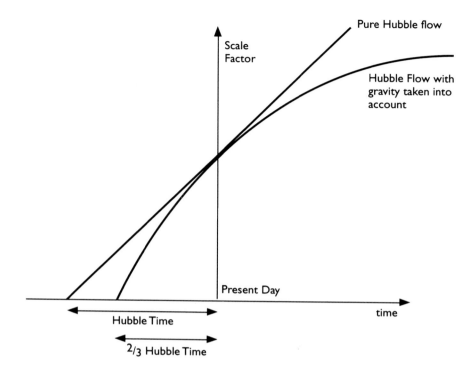

curved, open universe would be one which does not contain enough matter to halt the expansion. In our two dimensional sheet of paper analogy, this would correspond to a three dimensional saddle shape. The dividing line between the two, in which the universe contains just enough matter to halt the expansion after an infinite time has passed, is known as the flat case since the spacetime continuum would not be curved at all (see *figure 4.17*).

The various curvatures can be thought of as being similar to the curve of the Earth. Close by, we are unable to tell that the Earth is a sphere. We can detect its curvature only by looking at something a long way away and seeing it disappear over the horizon. In principle, astronomers should be able to observe objects at increasing distances and detect any overall curvature of the universe.

Figure 4.17
The curvature of spacetime

Negative curvature of the spacetime continuum produces a closed universe whose two dimensional analogy is a sphere. In a negatively curved universe the angles inside a triangle add up to more than one hundred and eighty degrees.

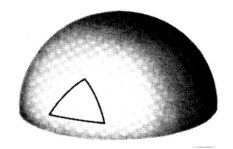

A flat universe, also known as a Euclidian universe, occurs at the boundary dividing open and closed universes. In a Euclidean universe the angles inside a triangle add up to exactly one hundred and eighty degrees.

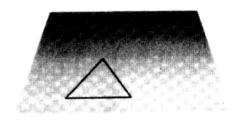

Positive curvature of the spacetime continuum produces an open universe whose two dimensional analogy is a saddle. In a positively curved universe the angles inside a triangle add up to less than one hundred and eighty degrees.

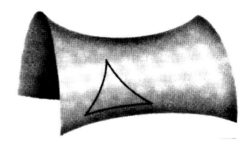

In the same way that the curvature of the two-dimensional sheet of paper took place through a 'higher' (third) dimension, so the curvature of the spacetime continuum takes place in a higher dimension than we are incapable of perceiving directly. In order to visualise the observational effects this would have, it is necessary to return to the paper sheet analogy. In this analogy you must imagine that you are a two dimensional being and as such you would not be able to perceive the curvature through the third dimension. In order to discover what you would observe in this situation imagine that the paper sheet can be

Figure 4.18
The observational effects of spacetime curvature

In a closed universe the density of galaxies will appear to decrease the further out into space one observes.

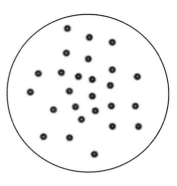

In a flat universe the density of galaxies will appear to remain constant the further out into space one observes.

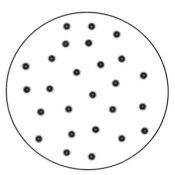

In an open universe the density of galaxies will appear to increase the further out into space one observes.

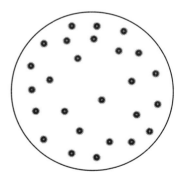

stretched or compressed like rubber and that it is covered in regularly spaced polka dots (see figure 4.18). As far as you can perceive, however, the spacetime continuum is flat. Forcing the hemisphere to become flat will show us the observational effects of the curvature.

In the negatively curved case, the edges would have to be stretched and so the distance between the polka dots would increase. Thus, the density of polka dots would decrease the further one looked out into space. In the case of a positively curved universe, the situation would be reversed. The edges of the saddle would have to be compressed in order to make it flat. Thus the polka dot density would increase with distance.

Obviously, in our universe the polka dots should be replaced by galaxies but the principle is the same. If our universe contains enough matter to halt the expansion and re-collapse, the density of galaxies should decrease the further into space we peer. If our universe does not contain enough matter to halt the expansion, the density of galaxies should increase the further into space we look.

A recurring theme to this chapter has been the complicating factors which are introduced when translating theory into practice. Unfortunately, this concept is no exception and the problem occurs because of look-back time. As we look at higher and higher redshifted objects, we are seeing them as they existed longer and longer ago. The universe, as we have been at pains to point out in this chapter, is expanding, thus when the light from these distant objects began its journey, the universe was indeed denser. It is therefore no surprise for us to observe an increase in the density of galaxies with increasing redshift. If the observed increase is less than predicted by the expansion of the universe alone then the curvature must be negative. On the other hand, if the observed increase is more than predicted by the expansion of the universe alone then the curvature must be positive. The exact expansion rate of the universe, as we have seen, is given by the Hubble constant which is uncertain by a factor of two!

Thus, cosmology remains one of the most fascinating, yet frustrating, branches of modern astronomy. The theories are in place but the data is trickling in. At this stage, all scenarios are still possible. Hope is on the horizon in the form of a new range of instruments, known as fibre optic spectrometers. These promise to turn the trickle into a flood because they are capable of recording hundreds of galaxy spectra simultaneously, whereas traditional spectrometers have only been able to record a single spectrum per observation. With the plethora of new spectra and, hence, redshifts soon to be flowing in from instruments such as the Two Degree Field Spectrometer on the Anglo-Australian Telescope in Australia and the Sloan Foundation's Digital Sky Survey in America, cosmologists should find themselves wallowing in tens of thousands of galaxy redshifts within the next few years.

Analysis will then allow the construction of the most detailed maps of universal structure yet. Through the use of the fundamental steps outlined in this chapter, this wealth of new data will hopefully allow the value of the Hubble constant to be pinned down much more precisely. Only when that has occurred will the cosmologists be able to really test their ideas about the universe.

5

Unconventional
interpretations of
redshift

Hubble consistently maintained that his work on redshift only gave 'apparent' velocities. When trying to analyse this statement it proves to be quite ambiguous. What exactly did Hubble mean? Was his phrase meant to show he believed in the expanding spacetime model or that redshifts were caused by a physical mechanism other than an expanding universe?

Although the expanding universe theories are the ones which are usually ascribed to explain the observable phenomena, especially the redshift, of objects within our universe, it is not the only explanation. A common thread, running through all previous chapters was that redshifts are created by some form of motion. Thanks to Einstein's principle of equivalence, even the gravitational redshift was explainable as an accelerated motion.

There are still a number of well-respected, professional astronomers who refuse to subscribe to the traditional interpretation of the cosmological redshift, namely one in which the spectral shift is caused by the expansion of the spacetime continuum. Instead, they have postulated that redshifts can be created by something other than motion. To differentiate this form of redshift from any other, it has become known as an intrinsic redshift. The idea is that some physical property or mechanism, which is intrinsic to the emitting object, systematically induces a redshift on the radiation. In the most extreme of cases, supporters of intrinsic redshifts refute that the universe is expanding at all!

Most astronomers who reject the standard view do not take quite such an extreme standpoint. They believe that, although the universe is expanding, thus causing a cosmological redshift, there are other mechanisms which can increase the redshift of objects. If this is the case, then everyone who assumes that redshifts are entirely cosmological in origin is overestimating the distance to celestial objects. The term cosmological distance is given to any estimate of a celestial object's distance which is based solely on the assumption that the redshift is cosmological in origin.

Proponents of the non-cosmological redshift ideas point to a number of astrophysical phenomena which appear to be at odds with the traditional view of the universe. Much of their evidence is based upon the statistical analysis of astronomical data. Other astronomers argue that there have not been enough observations of astronomical objects on which to perform such statistical analysis. Whatever the outcome, the redshift controversy is persistently in the background and will not go away until incontrovertible observational evidence one way or another is forthcoming.

Anomalous stellar redshifts

It is traditional for those who argue against the conventional redshift interpretation to base their arguments on the redshifts of galaxies and other extragalactic objects. Anomalous redshifts are not strictly confined to these and are also known to occur in the starlight from young, high mass stars. This, totally unexplained phenomenon, has been known since 1911 when it was announced by W.W. Campbell and called the K-Trumpler effect. The effect has been confirmed over and over again with published work from a number of different observing teams during almost every decade since!

Discrepant redshifts

By far the most evidence against traditional redshift interpretations has been accumulated for extragalactic objects. The previous chapter described how galaxies cluster together. The hierarchical structures of these conglomerations are such that a large dominant galaxy will be surrounded by a number of smaller companion galaxies. For example, the Andromeda galaxy in our local group has companion galaxies of M32, M33, NGC

205 and NGC 185. These objects are assumed to be in orbit around the larger galaxy and so, if spectra were taken, we would expect to see some displaying a blueshift whilst others displayed a redshift when compared to the spectrum of the Andromeda galaxy. This would arise naturally because the companions should be at random points in their orbits. Hence some would be moving away from us and some would be moving towards us with respect to M31. However, in reality, each one of the companions displays a redshift when compared with Andromeda!

Although it is unlikely, it is not beyond the realms of possibility that this result is simply a freak coincidence. If the study is extended, however, the same pattern begins to be noticed over and over again. In all, eighteen independent studies of this phenomenon would appear to have confirmed that companion galaxies generally exhibit higher redshifts than their dominant galaxies.

The mainstream astronomical community is often criticised for ignoring this data. Rudimentary explanations have been proposed although very little work has gone into the problem. Perhaps the best explanation conventional astronomy has to offer is that companion galaxies are falling into their dominant galaxy. We observe only the redshifting ones because the companions on the far side of the dominant galaxy, which would appear blueshifted, are obscured from our view because of intervening dust and gas (see figure 5.1).

Perhaps one of the strongest cases of discordant redshifts occurs in the famous galaxy grouping known as Stephan's Quintet. This association of five galaxies was first noticed by astronomer M.E. Stephan in 1877. The advent of modern telescopes and observational methods has turned it into one of the most controversial objects throughout the whole of space! As the name implies it is a collection of five galaxies, all of which

Figure 5.1
Collapsing galaxy systems

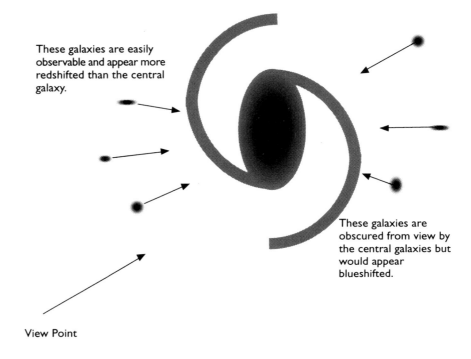

These galaxies are easily observable and appear more redshifted than the central galaxy.

These galaxies are obscured from view by the central galaxies but would appear blueshifted.

View Point

appear to be interacting with one another. Interacting galaxies are more usually pairs of galaxies. They are thought to mark the fate of many binary galaxy systems because as the galaxies orbit one another it is entirely likely that gravitational forces will cause them to eventually merge together. Interacting galaxies are observed to display long 'tails' of dust, gas and stars which have been thrown out of the merging galaxies. These are thought to be caused by the gravitational forces which raise tides on the galaxies in the same way as the Earth and Moon raise tides upon each other. These galactic tides, however, are so strong that they actually cause material to be thrown out behind each of the merging galaxies.

Several tails of this kind are visible in Stephan's Quintet, most notably from the galaxies NGC 7319, NGC 7320 and NGC 7318B. The remaining galaxies in the group are NGC 7318A and NGC 7317. The system looks perfectly acceptable until the galaxies' redshifts are taken into account. The measurements were made in 1961 by Geoffrey and Margaret Burbidge who converted their redshifts into separation velocities (from the Milky Way) for the presentation of their results. Most galaxies in the group (NGC 7319, NGC 7318A and NGC 7317) have a redshift which corresponds to a separation velocity of 6,700km/s. NGC 7318B, which is a highly peculiar object anyway, has a slightly lower figure of 5,700km/s. The big fly in the ointment, however, is NGC 7320 which was discovered to exhibit a redshift corresponding to a separation velocity of just 800km/s. This is a real puzzle because it clearly displays a tidal tail which means it is gravitationally merging with something. That something looks like the other galaxies in Stephan's Quintet but how can that be true if redshift is truly proportional to distance?

This perplexing galaxy system remains an unresolved problem in astronomy. Complicating the possible interpretations is NGC 7331, a nearby Sb spiral galaxy. This object appears much larger than the other galaxies in Stephan's Quintet and displays a redshift approximately equal to the low value given by NGC 7320. Investigation shows that the latter galaxy is a dwarf galaxy. NGC 7319 and NGC 7318A both look like disturbed spiral galaxies but are much smaller that NGC 7331. In fact, the system could be interpreted as a chance juxtaposition of the dwarf galaxy with the merging galaxies of Stephan 'Quartet' if it were not for the dwarf galaxy's obvious tidal tail. Some who believe that redshift cannot be relied upon to give distances believe that the different Hubble types of galaxy offer the key to the different redshifts displayed by galaxies.

Redshift and Hubble galaxy type

Galaxies appear to possess a myriad of different sizes and shapes. The shapes were classified in the previous chapter according to the Hubble tuning fork diagram. The sizes of individual galaxies, however, is a little more tricky. When two galaxies are seen in photographs and one is obviously bigger than the other does this mean that the bigger galaxy is truly the larger of the two or that the smaller one is just as big but seen from farther away? Traditionally it has always been assumed that galaxies, especially those of the same Hubble type, were all similar sizes and that their angular size has been an indication of how far away the galaxy exists. The redshifts of (apparently) smaller galaxies also appear to be higher thus lending support to the assumption that the galaxies are further away. Galaxies which are further away appear fainter to us than galaxies close by and so it becomes possible to create graphs illustrating the way in which the apparent magnitude of galaxies change with redshift.

When looking at the spiral galaxies of type Sc a discrepancy occurs. In the years following Hubble's original classification of galaxies, a number of sub-classifications have been added. One of these is to the Sc galaxies. Sc I galaxies are essentially the same as Sc galaxies except that the spiral arms are narrower and better defined. When their luminosities are plotted against redshift, it would appear that Sc I galaxies are much brighter than their siblings, the Sc galaxies *(see figure 5.2)*. When the diameters of the Sc I galaxies are calculated from their assumed redshift distances and their angular sizes, another problem occurs. The calculated sizes of Sc I galaxies become larger at greater redshift. In fact, the calculated sizes of these galaxies become so great that they eventually dwarf the Sc and Sc I galaxies in our local neighbourhood. They then go on to dwarf the Sb spiral galaxies (such as the Andromeda galaxy) which, according to study of the nearby galaxy groups, are much larger and brighter than Sc I galaxies. The situation is exemplified by Sc I galaxy NGC 309 in Cetus which, if it exists at its redshift distance, is so large that the nearby Sb spiral galaxy M81 in Ursa Major, which has traditionally been thought of as a large galaxy, would fit comfortably and very

Figure 5.2
Sc Galaxies

The black line represents the Hubble relation for Sc galaxies. The grey spots are data points for individual Sc I galaxies. This graph would suggest that Sc I galaxies are uniformly brighter than their Sc counterparts.

(Adapted from Arp and Block)

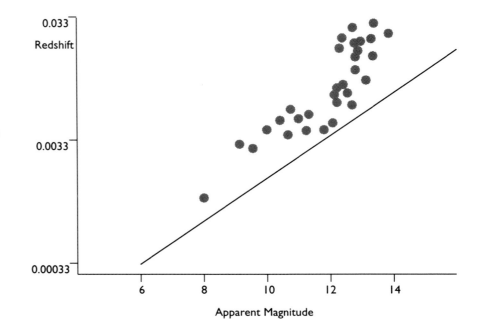

Figure 5.3
The size of NGC 309

This schematic representation shows the calculated size of NGC 309 (larger galaxy), based upon its redshift, in comparison with spiral galaxy M81 in Ursa Major (smaller galaxy). The size of M81 is well known because it is close to our galaxy and has been well studied.

unassumingly into a portion of one of NGC 309's spiral arms (see figure 5.3)!

An independent measure of the distance to these galaxies is needed in order to check the distances which are being derived from the redshifts. Astronomers Halton Arp and David Block used the Tully-Fisher method to do this and found some interesting results. The Tully-Fisher method for distance determination relies on the rotation of a spiral galaxy's disc to produce Doppler broadening in the galaxy's absorption lines. From analysis of this, the intrinsic luminosity of the galaxy can be estimated and the distance can then be calculated by equating the apparent magnitude of the galaxy to its actual luminosity. This distance should be approximately equal to the distance derived from the redshift if the redshift is, indeed, a good indicator of distance. In the case of Sc galaxies, the distance comparisons are good but in the case of the Sc I galaxies, the Tully-Fisher distances are smaller than the redshift distances by up to 100 million light years!

The final piece of information presented by Arp and Block to support their assertion that there is an excess redshift in Sc I galaxies, involves supernova rates. Supernovae come in two type, the first type occurs when a compact white dwarf star accretes matter from a giant companion star. This can trigger a catastrophic nuclear detonation which blows the white dwarf to bits. The second type occurs when a short-lived, high mass star reaches the end of its life cycle and its core collapses. Again, this produces a runaway nuclear explosion which completely destroys the star. It is in this second type of supernova which we are interested. They occur in the disc region of a spiral galaxy where the spiral arms are located. This is the site of star formation and the high-mass stars live such relatively short lives (typically a few million years) that they never leave the spiral arms.

The supernova rate per galaxy is very roughly one every century, based upon our observations of close by galaxies. If Sc I galaxies are much larger than the nearby galaxies, then their rates of supernovae explosions will be higher because they contain many more stars. If the size of NGC 309 is correct, it should average one supernova every three years. This is a rate which is simply not observed. If the galaxy were closer than its redshift distance, this problem would disappear because the galaxy would not be so big. In its place would be the problem of what causes the excess redshift?

In studying discordant redshifts, some astronomers began to notice another peculiar trend in the data. It appeared as if a disproportionate number of galaxies existed at certain redshifts! For reasons which defied conventional explanation, it seemed as if galaxies in groups contained redshift differences which were an integer multiple of 0.0002! This was an extraordinary result which was met with scorn and derision when it was first presented to the astronomical community. Now, after almost twenty years of study, the controversy has still not been resolved.

Quantized redshifts

The first person to bring the astronomical community's attention to this odd behaviour of data was an American astronomer named William Tifft. He was studying galaxies in the Coma cluster and investigating the way in which their luminosities depended upon their redshift. Instead of a random scatter, Tifft was astonished to see his data segregate itself, according to Hubble type, into different sloping bands. It was as if slightly different Hubble laws applied to the different types of galaxy! It was during these studies that Tifft began to notice a clumpiness to his data: as if galaxies were collecting around certain, preferred values of redshift.

This phenomenon, known as redshift quantization, is a perplexing one which was first noted in individual galaxies. Both M51, the Whirlpool galaxy and M31, the Andromeda galaxy were found by some research teams to possess velocity discontinuities. These occur when adjacent regions of the same galaxy are studied spectroscopically and found to show different redshifts. The discontinuities tend to occur most strongly in the nucleus of galaxies. Tifft then extended his work to pairs of galaxies. He presented data which showed that, in general, the redshift difference between any pair of galaxies in his sample deviated by a fixed multiple of his redshift quantization interval.

The first obvious interval was found to correspond to a Doppler velocity of about 72km/s. As later work continued, intervals of a half, a third and even a sixth of this value were shown to be possible too. This work on redshift quantization is of crucial importance to our understanding of the universe at large. If it is correct, then it means that physics, as we understand it, is not applicable to the largest scales present in the universe.

Some astronomers argue that there are no valid reasons why it should be. After all, we know that our classical laws of physics are not valid when trying to explain the microcosmic realm of the atom so why should they explain the realm of the galaxies? In their view there may be some physical mechanism which is negligible in our day to day scale but becomes important on the large scales present in the universe.

For example, modern cosmology is awash with hypothetical particles which are supposed to populate space in vast numbers. They go by a dizzying array of names, most of which seem to end in the suffix 'ino' and are required by certain, as yet unproven, cosmological theories. These exotic particles are highly attractive to the modern astronomer because they can potentially fill a great niche in our understanding of the cosmos.

It is currently thought that much more matter than can actually be seen is present in the universe. A property of this dark matter is that it is supposed to hardly interact at all with 'normal' matter, except through gravity. The fact that these hypothetical dark matter particles hardly interact with atoms on the scale of the world, or solar system, is obvious because, if they did, we would have noticed their effects by now. On the distance scales between galaxies, however, the story may be very different. Even something which only interacts with one atom over the distance of a light year will still produce over two million reactions between the Milky Way and the Andromeda galaxy. Thus, over the diameter of the solar system we may expect only one interaction if we are lucky. Even if it were to happen, we would never notice unless it occurred under our noses and then only if we were looking for it! Over the average distance to a galaxy, however, we can typically expect hundreds of millions of interactions. What if these hypothetical particles react in some way with photons of light? Perhaps they reduce the energy of the photons and induce a redshift?

It is a sobering thought, indeed, that on the large scale, our knowledge of physics may be woefully incomplete. It is one that astronomers Bruce Guthrie and William Napier, of the Royal Observatory Edinburgh, were forced to face when their investigations showed evidence that any galaxy, anywhere in the sky, has a redshift which falls into one of the predetermined intervals. Although they had initially set out to disprove the quantization results, they found that no matter what they did, the data stubbornly refused to give up its repetitious redshift structure.

As extraordinary as the existence of galaxies at preferred redshifts seems to be, it is not without astronomical precedence. In 1967, shortly after the discovery of quasars, those extraordinarily luminous active galaxies which exist at vast redshifts, it was pointed out by Geoffrey Burbidge and Margaret

Burbidge that they seemed to cluster around preferred redshifts. K.G. Karlsson showed that quasars' redshifts appeared to follow a precise mathematical formula. This was apparently disproven by work conducted at the Cambridge Institute of Astronomy which included a much larger sample of faint quasars into the analysis. The claims and counter-claims for and against redshift quantization still reverberate in astronomical circles. If the quantization really is proven it would mean the quasars exist at redshifts in a steady series of steps just as electron energy levels do around atomic nuclei *(see figure 5.4)!* In fact, the quantization intervals of redshifts can be listed in ascending order from nearby galaxies, to those in clusters and groups, out to the quasars.

Figure 5.4
Redshift quantization

The evidence for redshift quantization postulates that every galaxy exists in a predefined redshift. On this diagram the Milky Way is the galaxy in the centre and the concentric circles mark the redshift quantization intervals of 0.0002.

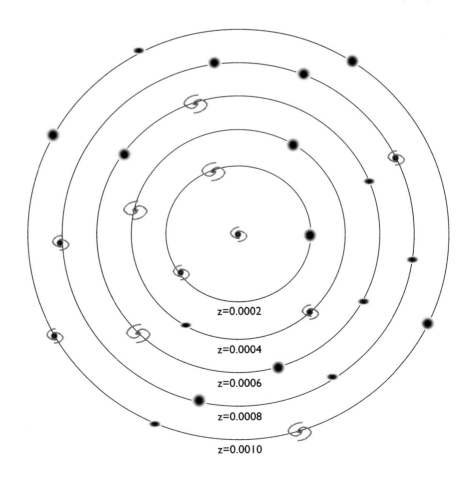

z=0.0002

z=0.0004

z=0.0006

z=0.0008

z=0.0010

Figure 5.5
Binary galaxies

The left hand binary galaxies are widely spaced when projected onto the sky plane. This makes them easy to overlook when identifying associated pairs. Their velocities are largely tangential and so they produce very low redshifts. If these are overlooked, it can lead to an apparent lack of low redshifts. The right hand binary galaxies are easy to spot as a pair because they are close to one another. Their velocities are largely radial and so they give large redshifts. This inadvertent exclusion of low redshifts may make the data look quantized by causing a peak at higher redshifts.

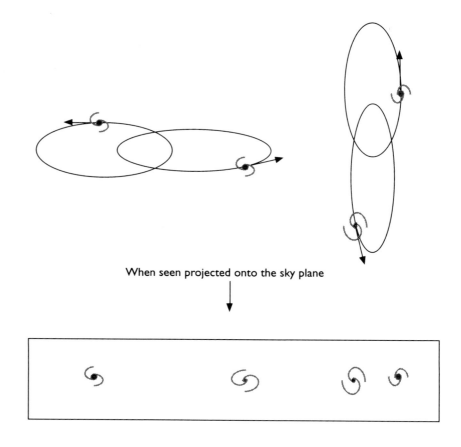

When seen projected onto the sky plane

Of course, the majority of astronomers are not convinced by the quantization evidence. Many believe that it is a statistical fluke. For example, if binary galaxies are being studied then they will be identified by their proximity to each other on the sky plane. The tendency will be to miss the widely spaced binaries whose components are at extremes in their orbits (see figure 5.5). This means that the sampling of binary galaxies will be incomplete and inaccuracies will be introduced into the data.

The argument about the reality of this phenomenon does not look as if it will be resolved in the near future. The quality of the data is of vital importance to the debate. Has inaccurate data produced the quantization peaks, which have then been elevated by statistics to an unreal level of importance? Or does inaccurate data hide the quantization peaks from view? It is no

Figure 5.6
Pencil beam surveys

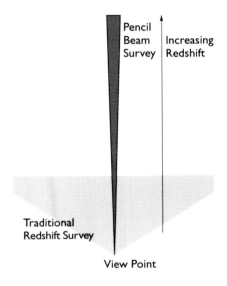

Pencil Beam Survey

Increasing Redshift

Traditional Redshift Survey

View Point

distortion of the facts to state that more papers have been published which claim to show the quantization than those which claim to disprove it. Most papers, however, simply assume that quantization is an unreal effect in the data and ignore it.

Periodic redshifts due to clustering

In recent years a form of periodic redshift, which is totally separate from the quantization effect, has also been observed. Traditionally, redshift surveys have involved scanning large areas of sky for any galaxy down to a preset brightness limit. This gives us an idea about how the galaxies are distributed around us. A different approach, known as a pencil-beam survey, studies a small region of the sky down to an incredibly faint limit. Thus it probes much further into the universe than a low depth redshift survey does. Using this information, it is possible to get a feel for the way in which galaxies are distributed at great distances from us (see figure 5.6).

In the late 1980s, two pencil-beam surveys were conducted independently of one another. The first, led by Thomas J. Broadhurst from the University of Durham looked at the northern hemisphere, whilst the other, led by David Koo from the University of California, Santa Cruz, surveyed a small region of the southern hemisphere. The data collected by both teams was fascinating. Instead of a smooth distribution of galaxies with increasing distance, they discovered that galaxies bunch together at apparently regular intervals. This was an unexpected result to say the least. The two teams pooled their data and presented it in a dramatic bar chart which showed the number of galaxies which existed at each redshift (see figure 5.7).

The results were converted into distances using a Hubble constant of 100 km/s/Mpc and the gaps in between the galaxy clumps were discovered to all be about 400 million light years. To give some idea of the scale upon which the survey was

Figure 5.7
Periodic redshift due to clustering

Adapted from Broadhurst et al 1990

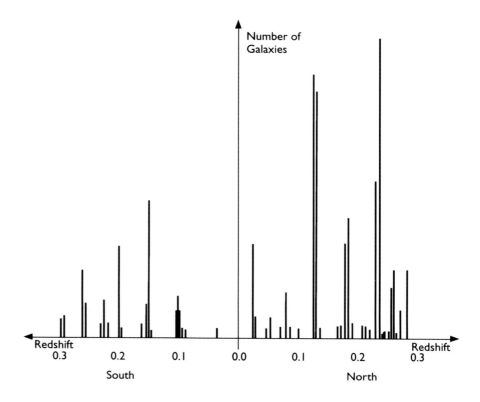

undertaken, the largest structure in the known universe is the great wall of galaxies mentioned in the previous chapter. This vast structure, which is far wider than it is deep, showed up on the northern hemisphere pencil-beam survey as only the first, nearest, clump. Behind it were many more clumps, all of which could be 'great walls' in their own right.

The only way in which the current data can be explained is if the clumpy structures observed represent the 'front' and 'back' surfaces of superclusters which are surrounding vast empty voids (see *figure 5.8*). If this is the case, then adjacent pencil-beam surveys would show different patterns to the periodicity and by combining the surveys, the expected, smooth, distribution of galaxies throughout space should be found (unless the ideas about quantization of redshift are true). It is necessary to point

Figure 5.8
**Three dimensional structure
of the universe**

Periodic structure in the redshift
data from pencil beam surveys is
understandable if superclusters of
galaxies are distributed around the
surface of vast spherical voids.

View Point

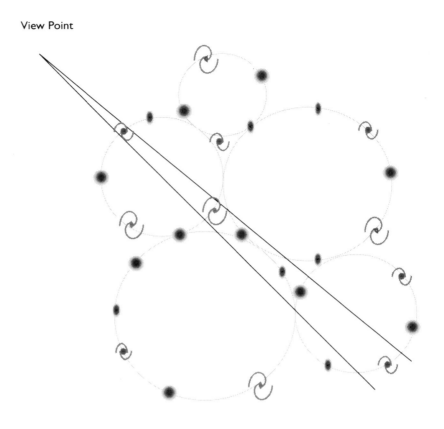

out here, however, that the reason the periodic redshifts are
treated as a separate problem from the quantized redshifts is
because the clumping takes place on vastly larger scales. For
periodic redshifts, the redshift intervals are very approximately
0.1. For quantized redshifts, the redshift intervals are
approximately 0.0002.

As stated in the previous chapter, astronomy is facing the
welcome flood of new redshift data from instruments such as
the Anglo-Australian Two Degree Field project and the
American Sloan Digital Sky Survey. This will really help to show
us whether periodic redshifts are a breakdown in our
understanding of the universe or the concrete proof that
superclusters surround vast bubble-shaped voids.

Another key area of the redshift debate has been over the nature of quasars – those enigmatic beacons which shine in space but, because of their small angular size, refuse to give up their secrets easily.

Superluminal motion

In the 1970s a technique known as Very Long Baseline Interferometry, often abbreviated to VLBI, was introduced into the mainstream of observational techniques. Using this methodology it became possible for observatories on different continents to study objects at radio wavelengths and then combine their results to produce data with unprecedented angular resolution. In certain cases it became possible to show features which were as small as a thousandth of an arcsecond across. It became an obvious choice of technique with which to study the objects known as quasars.

Strictly speaking, a quasar is the term given to a quasi-stellar object (QSO) which emits radio radiation. This is because the name, quasar, is actually a rather clumsy concatenation of the term: quasi-stellar radio source. This division of terms proved rather unwieldy for most astronomers since quasars and QSOs tend to be principally identified by their optical spectra, which give little clue to their radio emission. Hence the term quasar has come to be used to reference any type of QSO. If qualification is needed, then a phrase is appended stating whether it is radio loud or radio quiet.

The reason the radio loud quasars became a natural choice for study was because, in even the largest telescopes, they only ever appeared as star-like point sources (hence the name quasi-stellar object). With the milliarcsecond resolution made possible by VLBI, however, it was hoped that some inner detail would finally be found.

Mathematical note 5.1

Faster than light motion

If an active galaxy expels a jet of material, its passage away from the galaxy can be observed and its transverse velocity calculated:

$$v_t = \frac{d}{t_2 - t_1}$$

where
v_t = transverse velocity
d = distance moved in between observations
t_1 = time of first observation
t_2 = time of second observation

The transverse velocity must be interpreted as a lower limit for the true velocity of the jet. This is because the jet will almost certainly be travelling in a direction which is not in the plane of the sky. Thus the transverse velocity is a projection of the true velocity onto the sky plane and will only represent a component of the motion:

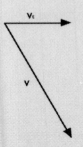

When this calculation is made for jets in many active galaxies, the transverse component of velocity is calculated to be greater than the speed of light! This is the phenomenon known as superluminal motion.

When observations commenced, it was with great excitement that many quasars appeared as if they were ejecting blobs of material which glowed at radio wavelengths. This idea of ejection became confirmed several years afterwards when follow-up observations showed that the 'blobs' had moved. Excitement soon turned to bewilderment as attempts to calculate how fast the blobs of ejecta were moving yielded the unbelievable result that many were being shot out from hearts of quasars at velocities greater than the speed of light (see mathematical note 5.1)! In fact, over half the quasars studied in this way have been observed to display faster than light expansion. It became termed: superluminal motion.

This apparent contravention of the laws of special relativity is not actually as serious as it at first sounds. In fact, there are at least two possible explanations. The first, which happens to be the accepted one, is called the aligned beam model. This postulates that the material which we can see being ejected is actually being flung towards us at a velocity which is a large proportion of the velocity of light. This idea relies on the fact that the moving material is travelling at an angle which is slightly off-set from the line of sight. If this geometry were present, then the distance travelled by the radiation given off when the blob was ejected, would not vastly exceed the distance travelled by the 'blob' itself. The ejected material would continue to emit radiation which would stream to Earth in a continual flood. The radiation would have a substantial head start, however, on that being emitted by a clump of material which was moving at right angles to our line of sight.

The problems of gauging distances were highlighted in the last chapter when projection onto the celestial sphere was discussed. In the case of superluminal motion, projection of the forward moving clump onto the celestial sphere makes it appear as if the ejected matter is moving at right angles to our line of sight. When this radiation is emitted by the clump we vastly

Mathematical note 5.2

The geometrical explanation of superluminal motion

The phenomenon of superluminal motion can be shown to be an optical illusion caused by the emission of the jet, at relativistic velocities, at an angle very close to that of the line of sight.

A — ejection takes place at time t_1

Light from the ejection begins its journey at time t_1

θ — ejecta moves with velocity V

Light from the ejection has reached this B position at time t_2

C — ejecta at time t_3

Light from the ejection has reached this D position at time t_3

towards earth

The distance between A and C is:

$$AC = V(t_3 - t_1)$$

Therefore the distance between A and B is given by trigonometry as:

$$AB = V(t_3 - t_1)\cos\theta$$

The distance between A and D is:

$$AD = c(t_3 - t_1)$$

The distance between B and D is:

$$BD = AD - AB = (t_3 - t_1)(c - V\cos\theta)$$

The time it takes light to travel this distance is:

$$(t_3 - t_2) = \frac{BD}{c} = \frac{(t_3 - t_1)(c - V\cos\theta)}{c}$$

In this time, the ejected material appears to have moved sideways from B to C, the distance of which is:

$$BC = AC\sin\theta = (t_3 - t_1)V\sin\theta$$

Thus, the transverse velocity of the blob is given by:

$$v_t = \frac{AC\sin\theta}{(t_3 - t_2)}$$

which simplifies by substitution to:

$$v_t = \frac{V\sin\theta}{1 - \dfrac{V\cos\theta}{c}}$$

For ejection angles close to the line of sight, $\cos\theta$ tends to 1 and although $\sin\theta$ tends to 0, the effect is that as the velocity of the jet tends to the speed of light, the apparent transverse velocity exceeds the speed of light.

overestimate its velocity. This is because, although the clump appears to be moving in the plane of the sky, it is actually heading towards us. Its radiation, therefore, is given an enormous head start and we derive velocities in excess of the speed of light *(see mathematical note 5.2)*.

Arguments against this idea rely on statistics and arise because these superluminal jets appear in over half of the studied radio emitting quasars. It seems very strange indeed that such a large proportion of the studied objects should be throwing clumps of gaseous matter in our direction when, statistically, only one in every fifty jets should be aligned in such a way that superluminal motion is observed.

Although this seems like a pretty convincing argument against the beamed radiation model, its proponents can counterattack with some statistics of their own. Radio quiet quasars actually outnumber the radio loud variety by about 100 to 1. A central question in astronomy right now is whether or not a radio quiet quasar has a radio loud component which simply cannot be detected. This question is at the heart of the debate concerning the nature of superluminal motion. It also reintroduces us to the phenomenon of Doppler boosting which was discussed in Chapter Two.

We assume that the radio jets are rather faint and therefore very difficult to observe. Those jets which are tilted towards us will be Doppler boosted to appear much more luminous. Thus they will be easier to observe. The more they are tilted towards our line of sight, the greater the Doppler boosting and so the more likely we are to notice them. Obviously, the closer they are to our line of sight, the more susceptible they are to displaying superluminal motion. Thus, runs the argument, it is no surprise that a disproportionate number of the jets we observe show superluminal motion.

Sceptics point out that this explanation is rather convoluted and that a very rudimentary analysis of the mathematics which describe these jets allows a much easier solution to be proposed. The phenomenon of superluminal motion will disappear completely if these quasars are actually much closer than the distance indicated by their cosmological redshifts *(see mathematical note 5.3)*. Thus, there is no need for recourse into the ideas of Doppler boosting and line of sight effects.

Currently one of the best examples of a superluminal jet belongs to quasar 3C273 which can actually be viewed in optical light. 3C273 is a famous object because it was the first quasar

---

Mathematical note 5.3

Distance explanation of superluminal motion

In order to calculate the distance moved by the material which is ejected by the galaxy, a knowledge of the galaxy's distance from us is necessary. This is because, from Earth, only the angular distance of the ejecta's motion is measurable. In order to turn this into a linear measurement, the galaxy's distance must be used:

$$S = D\theta$$

where
S = linear separation of ejecta from galaxy
D = Distance of galaxy
θ = angular separation of ejecta from galaxy

After a set time, t, the separation has increased such that:

$$S + s = D(\theta + h)$$

where
s = extra linear separation
h = extra angular separation

Thus, the linear velocity of the ejecta, V, is given by:

$$V = \frac{s}{t} = D\frac{h}{t}$$

Since both h and t are measurable quantities they cannot vary. The distance, D, however, can only be estimated in most cases. If the distance estimates are too high, the linear separation velocity will also be too high. Thus, one way to reduce the velocity of the expansion is to assume that the galaxy distance estimates are too high. If D is reduced, V becomes smaller as well. Superluminal motion may, therefore, be an artefact created because the distance estimates of galaxies are too high.

to ever be discovered. It lies in the direction of Virgo and is projected into the heart of the Virgo cluster of galaxies. The redshift of the Virgo cluster is known quite accurately to be 0.003 whereas 3C273 displays the larger redshift of 0.158. Images of the jet, taken by the Hubble Space Telescope, have been interpreted by a team of astronomers, led by Robert C. Thompson, to indicate that the jet is being ejected in a direction almost parallel to the plane of the sky. In other words, at about ninety degrees to our line of sight, which is a totally unacceptable angle for superluminal motion to take place. If this result is confirmed by follow-up observation using the Hubble's repaired optics, it would appear to offer strong evidence that 3C273 exists more closely to us. The quasar may, in fact, be so close, that it would actually be a member of the Virgo cluster!

Quasar pairs

A number of astronomers seemed very surprised indeed to discover examples of quasar pairs. As the name implies, these are two quasars which lie close together on the sky plane. They divide naturally into two categories, those pairs which have the same redshift and those pairs which display discordant values of redshift from one another.

Quasar pairs with the same redshift are now understood to be gravitationally lensed. This is a consequence of the way in which spacetime is curved by the presence of matter. Double, triple and even quadruple images of the same quasar can be produced when the light is bent around an intervening galaxy (see figure 5.9). Periodically, quasars vary in their light output. When this happens in a gravitationally lensed system, one image will change its appearance before the other. This is because the light from different images will be reaching us along different paths which are of slightly different lengths (see figure 5.10). In general, the magnitude of these time differences will vary in proportion to the distance of the gravitationally lensed quasar. If

Figure 5.9
Gravitational lenses

Light from the quasar is deflected by the gravitational field of the intervening galaxy. This means that when it is viewed from Earth, the light rays seem to show multiple quasar images on the sky plane. All the quasar mirages will have identical spectra.

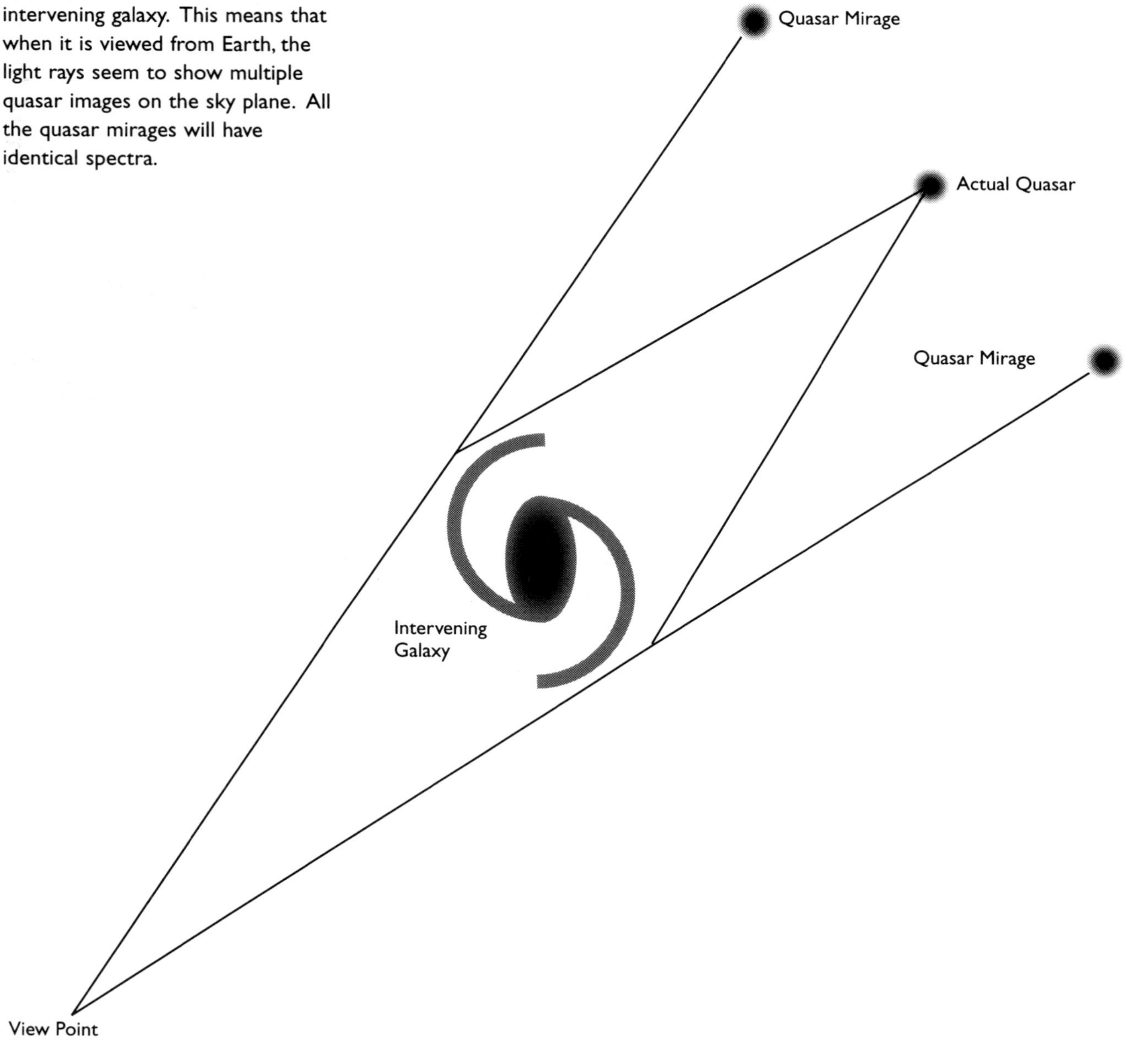

Figure 5.10
Time delays in gravitational lenses

In a mis-aligned gravitational lens system one path length is always greater than the other. In this system, a brightening in the actual quasar will be mirrored in quasar B first and, shortly afterwards, in quasar A.

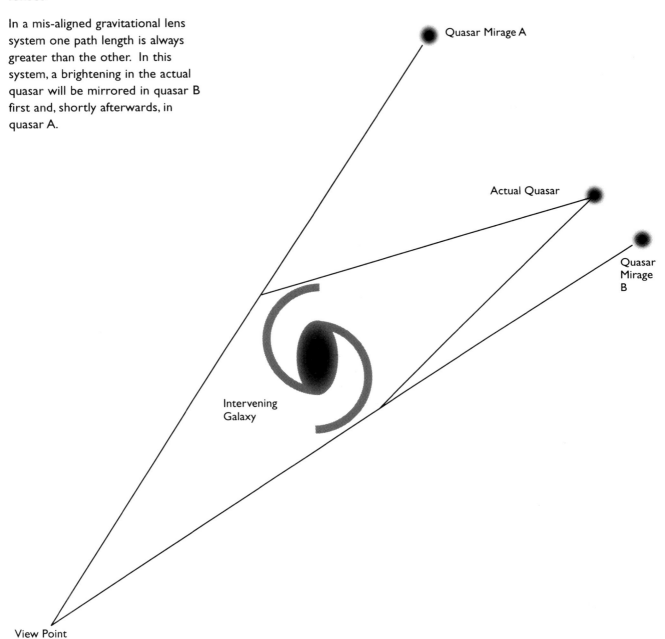

Quasar Mirage A

Actual Quasar

Quasar Mirage B

Intervening Galaxy

View Point

it is close by, the time delay will be shorter than if the quasar lies at a vast distance. Such timing experiments have been analysed by Arnon Dar of the Technion-Israel Institute of Technology, Israel. He showed that the evidence collected on the quasar Q0957+561, which is the only object to have had its image delay timed, strongly points towards a cosmological distance for this quasar. Although it has also bolstered belief in all the other lensed systems, some astronomers still hold out from subscribing to the party line on these objects and point to a famous lens candidate known as the Einstein Cross as their evidence.

The angular separation of the lensed images allows the mass of the intervening galaxy to be calculated. This has been performed for a collection of four images of the same quasar known as the Einstein Cross. Further measurements of this system show that the intervening galaxy's mass must be contained within a relatively small volume and this has led to an anomalously high figure for the galaxy's expected brightness. In fact, the galaxy should be intrinsically more luminous than a quasar would be at that distance! This seemingly absurd result has led Halton Arp to propose that the Einstein Cross is not a gravitational lens but an ejection event similar to the Markarian 205 and NGC 4319 system (which we will soon discuss).

Quasar pairs with different redshifts are still a matter of debate. There are two possibilities, the first is that they are simply chance alignments. This would mean that they exist at their redshift distances, are nothing to do with one another and simply appear to be close together. The possibility of chance alignment is hotly debated. Although galaxy pairs are also known, galaxies are far more numerous than quasars. Galaxies are also known to exist in clusters, thus forcing them closer together. If the less numerous quasars are assumed to be scattered randomly throughout space then the chances of a close alignment become very small (see *mathematical note 5.4*). The second idea is that they are actually physically associated

Mathematical note 5.4

Chances of quasar pairs

If the total number of quasars found in a survey of A square degrees of sky is, N, then the average number of quasars per square degree, n, is:

$$n = \frac{N}{A}$$

Having found one quasar, the probability, p, of another falling within an angular separation of x degrees by random chance is then:

$$p = \frac{\pi x^2 N}{A}$$

Using conventional values for the average number of quasars per square degree, leads to very low probabilities of finding quasar pairs. Yet, there are several examples of quasar pairs dotted about the sky. There are even triple and multiple quasars! Could it be that our quasar surveys are incomplete and underestimate their total number, or is this strong statistical evidence against quasars being randomly scattered throughout space?

objects. If quasars can exist in pairs, then the statistical concerns of their proximity vanishes. In its place, however, the fact that redshifts are not entirely cosmological in origin must be taken into account.

Quasar-galaxy links

The idea of quasars being somehow associated with galaxies, i.e. existing at the same distance, and yet showing an excess redshift is perplexing. Just three years after the discovery of quasars, both Halton Arp and, independently Fred Hoyle working with Geoffrey Burbidge, noticed that, for some peculiar reason, quasars seemed to cluster around the regions of sky near to bright galaxies. For example, the elliptical galaxy NGC 3842 holds court to three quasars, all of which appear very close in projection onto the sky plane. Given the number of quasars which have been observed to exist, the chances of three quasars falling close to a bright galaxy was estimated to be about one in a million! To make the statistics even worse, there are another three systems, NGC 622, NGC 470 and NGC 1073 which all have two or three quasars in close proximity to them. There are other galaxies with single quasars close by as well. To the astronomers who became intrigued by these systems, they presented good statistical evidence that something was wrong with the interpretation that quasars were at cosmological distances. They argued that the quasars were somehow associated with the galaxies they surrounded, perhaps even having been created by them.

The debate took a interesting twist in 1971 when Arp presented an image of galaxy NGC 4319 and the close by quasar-like object Markarian 205. These two objects both have published redshifts of 1,700km/s and 21,000km/s. It was assumed that Markarian 205 was at cosmological distances and that its proximity to NGC 4319 was simply a coincidental alignment. Arp's 1971 photograph showed that these objects appeared to be connected. This was such an outrageous view that most were highly sceptical. Other images, by different research teams, confirmed the observations and deepened the mystery. Many astronomers felt that the 'bridge' between Markarian 205 and NGC 4319 was a oddity produced by the way photographic film responds to light. With the advent of the charge coupled device (CCD) the pair were re-examined. A CCD is an electronic camera which acts like a photon counter. It will precisely count almost every photon which strikes its

Plate 5.1
NGC 4319 and Markarian 205

The anomalous bridge of material is arrowed.

(Halton Arp)

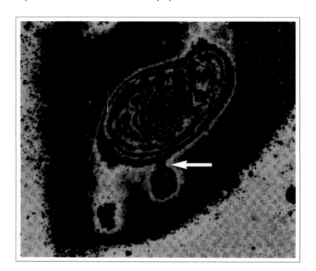

Mathematical note 5.5

The addition of redshifts

If one or more redshift is present in an astronomical object, the resulting redshift is given by the equation:

$$(1 + z_{tot}) = (1 + z_a)(1 + z_b)$$

where
z_{tot}=*total redshift*
z_a=*redshift caused by mechanism 'a'*
z_b=*redshift caused by mechanism 'b'*

At small redshifts this equation is well approximated by:

$$z_{tot} = z_a + z_b$$

detection circuit and by having these detectors rigged into a two dimensional array, an image can be taken. When NGC 4319 was imaged with these new detectors, which were not as fickle as photographic plates, instead of the bridge disappearing, the link to Markarian 205 showed up once again! *(see plate 5.1)*.

If the bridge is real then a possible explanation is that Markarian 205 is an object which has been violently expelled by NGC 4319. From our perspective, the 'quasar' has been thrown out backwards at a huge velocity and so added to its cosmological redshift is an enormous Doppler redshift *(see mathematical note 5.5)*. By interpreting the redshift solely as cosmological in origin, the distance to Markarian 205 has been grossly over-rated.

There are other reported links between galaxies and quasars. It would seem highly unlikely, however, that all can be explained by the ejection hypothesis because, statistically, half of the associated quasars should be ejected towards us. In this circumstance we would expect to see quasar-like objects with lower redshifts than their associated galaxies and blueshifts. This has never been observed and the subject remains an unexplained area of great controversy.

In 1980 an explanation for the numbers of quasars which appear close to galaxies was proposed by Claude Canizares. It relies on the phenomenon of microlensing, in which the light from faint quasars, which would not normally be noticed, is amplified by an intervening star in the halo of the galaxy acting as a tiny gravitational lens. Although it gives the quasar a boost in brightness, the gravitational field of the star is not great enough to split the quasar into separate images and so it remains as a single, albeit brighter, quasar (see figure 5.11 and plate 5.2). It is an attractive idea but as yet remains an unproven one on statistical grounds. Opponents to this hypothesis claim that there are simply not enough faint quasars to account for all the instances of accidental galaxy-quasar associations. Supporters of the idea claim there are enough quasars but we may simply not have discovered them yet!

Tired light

If the redshifts of distant galaxies and quasars are not cosmological in origin, then what causes them? The idea that electromagnetic radiation somehow loses energy as it races across the heavens has long been a popular alternative to the models in which the universe is expanding. As the radiation loses energy, so it becomes increasingly redshifted. This concept has become known as tired light. Perhaps its biggest claim to fame, and the reason it still persists today, is that it could successfully predict the temperature of the cosmic background

Figure 5.11
Gravitational microlensing

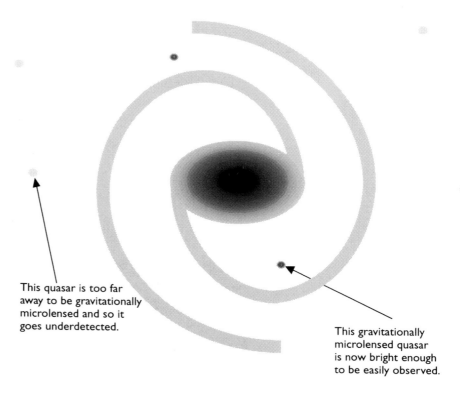

This quasar is too far away to be gravitationally microlensed and so it goes underdetected.

This gravitationally microlensed quasar is now bright enough to be easily observed.

Plate 5.2
Quasars around NGC 1073

Arrows point to three quasars.

(Halton Arp)

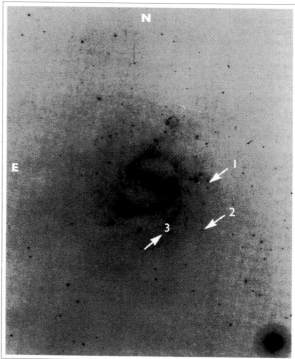

radiation to be 2.8 Kelvin whilst the big bang models were still predicting temperatures as high as 50 Kelvin! As far back as 1926, long before the ideas of the big bang or tired light were ever conceived, the English astronomer Sir Arthur Eddington calculated that the temperature of space should be approximately 3 Kelvin. Even today, some astronomers maintain that the background radiation temperature can be closely predicted by considering only the ambient starlight.

Ultimately, the tired light theory fails in one major way, which is that it is not an extension of current physics. Instead, it is an ad hoc addition: a 'what if' hypothesis which fitted the facts without really giving any good reason for them. In this case, why should light lose energy just because it travels through space? As such it joins the ranks of other theories such as the luminiferous aether which became popular but were subsequently proved fallacious.

The Wolf effect

In the last two decades, the work of physicist, Emil Wolf, has been hinting at a tantalising alternative to the three ways that a redshift can be produced and have been explained in this book. The new effect was discovered whilst studying lasers and other mechanisms which beam radiation in a preferential direction. Lasers are excellent examples of this, as are synchrotron sources in which emission is produced by electrons spiralling around magnetic field lines.

The Wolf Effect occurs when two sources are in close proximity and are emitting radiation of very similar wavelength. Wolf found that the radiation from the sources can interact and cause a wavelength shift. The precise value of the wavelength shift depends upon the angle from which the radiation is viewed. When viewed from a direct, 'head-on' vantage point the radiation is redshifted but at other angle the radiation can be blueshifted (see figure 5.12).

Figure 5.12
The Wolf effect

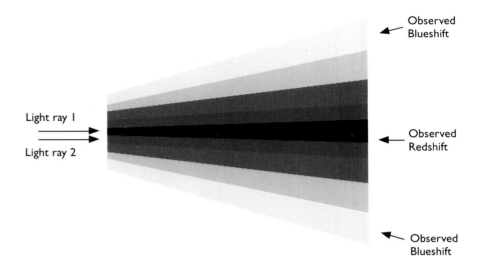

Emil Wolf and his colleagues have extended the idea from just two radiating sources to a whole collection of emitters such as would be found in a plasma cloud. They found that the shape of the emission lines could be redshifted effectively and that their shape was very similar to that of a Doppler shift. Their work may have a very important application to the jets seen in active galaxies which are beamed objects where a major source of radiation appears to be synchrotron in origin. The work may also translate into galaxy redshifts too, perhaps providing a mechanism by which excess redshifts can be generated.

The age interpretation

A radical alternative to conventional cosmological redshift interpretation has been presented by Halton Arp and colleagues. It is their contention that the redshift is an intrinsic property of matter which is imprinted upon the photons of radiation it subsequently emits. They believe that when matter is created it is brought into being with a large intrinsic redshift but that, as it ages, the redshift gradually diminishes.

The theoretical basis for Arp's assertions relies on the removal of a basic assumption in the standard model of the universe which was derived by Alexi Friedmann in 1922. Friedmann assumed that the mass of elementary, sub-atomic particles could never vary. Particles then communicate their mass with other particles via the exchange of ephemeral particles. These are known as gravitons and can never be directly observed, although their effects, the creation of mass and gravity, can be felt. Arp has taken as his working model that the number of these gravitons which can be exchanged actually defines the mass of the particle. The number of gravitons which can be exchanged, depends upon the number of other particles which fall within the newly created particle's horizon. Horizons were discussed in Chapter Four and for each subatomic particle, the horizon grows at the speed of light. Hence the mass of a particle changes with time as it finds itself able to communicate with more and more surrounding particles.

In Chapter One it was explained how radiation was absorbed and released from atoms when electrons jump up and down between the quantum levels. The precise amount of energy absorbed or emitted actually depends upon the mass of an electron which is traditionally thought to be a constant at 9.109×10^{-31} Kg. If Arp is right and the mass of elementary particles, such as electrons, does increase with time, then a redshift would naturally arise from younger objects. This is because the electrons would be less massive than their counterparts in our galaxy and, as a consequence, would emit less energy when jumping between atomic energy levels. Thus the radiation would have a larger wavelength and appear to be redshifted.

Look-back time is the key to understanding why we see distant galaxies exhibiting redshifts. We are seeing them at different stages in their evolution, depending upon exactly how far away they are located in space. A distant galaxy will appear younger than one which is close-by because its light has had to travel a longer distance to get to us.

This age interpretation for redshift has some dramatic consequences. For instance, Arp has concluded that time must have passed more slowly in the past when galaxies were less massive. In this view of the universe, space is static and matter can be created spontaneously by the rearrangement of energy in our universe. As the matter ages so its mass increases and its redshift diminishes.

Based on this work, Arp has predicted a redshift-distance law which appears to mimic the observed redshift-distance relationship to much greater accuracy than the Hubble law. It hinges on the fact that the redshift-distance relationship is not linear which is a conclusion other astronomers have also reached independently.

Non-linear Hubble laws

In 1925, shortly after the redshifts of distant galaxies had become firmly established, K. Lundmark wrote in Monthly Notices of The Royal Astronomical Society that their radial velocities (as indicated by the Doppler interpretation of their redshifts) might be represented by a quadratic equation which relied on the square of the object's distance (see *mathematical note 5.6*). Hubble's famous papers on the subject appeared to settle the debate once and for all by showing that Humason's observations could best be fitted with a straight line as shown in the last chapter. The linear Hubble law has been championed by most astronomers ever since but in recent decades, as more and more redshift data has become available, some astronomers are beginning to doubt the original interpretation. Any hint that perhaps galaxies do not fall into a uniformly linear distance scale based upon their redshifts has been met with quite severe

---

Mathematical note 5.6

Quadratic Hubble laws

Lundmark assumed that the separation velocities of distant galaxies would be described by a relationship which relied on distance in the following way:

$$V = A + HD + BD^2$$

where
V = separation velocity
H = Hubble Constant
D = distance
A and B = fitting parameters

Hubble's work showed that over the distance he considered, the separation velocity-distance relationship was fitted by:

$$V = HD$$

Some of the astronomers who have access to large amounts of cosmological redshift data, are now beginning to wonder if the Lundmark law is a better equation over very large distances. Hubble's law works close by because it is like a tangent to the Lundmark law:

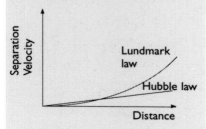

If the Lundmark law is correct, the big bang theory of the universe's creation is almost certainly wrong.

opposition. The anomalies have been ascribed to so-called selection effects which occur when a subset of galaxies are systematically left out of an analysis. Selection effects are usually unconsciously perpetrated and typically involve missing out a number of galaxies because they are too faint to be detected.

Recently, however, the realisation that streaming motions are present in the universe, i.e. clusters of galaxies being gravitationally attracted to other clusters of galaxies, has given traditionalists a powerful theoretical reason to explain why deviations from the linear Hubble law might occur. From studies of the brightest galaxies in nearby clusters, Sydney van den Burgh, estimated that deviations of up to twenty per cent are possible out to a redshift of 40,000 km/s. Even these considerations do not extend to the very large scales and the Hubble flow is still assumed to follow its linear course over vast distances. An increasing number of astronomers are pointing to data which appears to deviate strongly from the predictions of the Hubble law.

I.E. Segal of the Massachusetts Institute of Technology advocates a theory known as chronometric cosmology. It relies on a static cosmos which is analogous to a fixed sphere. It is the type of universe, in fact, which Einstein developed from his equations of general relativity before he learned of the redshifts possessed by galaxies. Chronometric cosmology predicts a Lundmark law, i.e. a quadratic equation, for redshift versus distance graphs. In an analysis of galaxies observed by the Infrared Astronomical Satellite (IRAS), Segal and his colleague, J.F. Nicoll of the Institute for Defences Analyses, Alexandria, Virginia, showed that the way in which the apparent magnitudes of these IRAS galaxies depended upon their redshift was correctly predicted by the Lundmark law but incorrectly predicted by the Hubble law.

The supporters of chronometric cosmology claim that its roots are based in traditional physics. In many ways it is to the tired light theory what photon theory was to Newton's light corpuscles: the same basic idea but steered in a subtly but importantly different direction. Its proponents build the theory's foundations on the notion that physics has three fundamental parameters which define its various regimes. For example, the regime of special relativity is reached when velocities approach the fundamental speed of physics: the speed of light. The realm of quantum theory is reached when the masses of objects approach another fundamental boundary denoted by the Planck constant. When the quantum theory failed to solve certain problems concerning ultraviolet light, one of the theory's greatest contributors, Werner Heisenberg, proposed that physics required a third fundamental parameter: that of length. It was Heisenberg's notion that the length should be of microscopic dimensions but he died before ever finding a convincing solution to this problem. The chronometric cosmologists, however, have resurrected this idea and shown that the fundamental length is not small but is very large instead. In fact, it is so large that it represents the curvature of the universe. Studies conducted by Segal and Nicoll estimate that this fundamental length is 160 million parsecs with an uncertainty of twenty five per cent either way.

In this theory the redshift is brought about because of the radiation's passage through the curved spacetime continuum. In essence, a time-dependent evolution to successively lower energies is imposed upon the radiation and it is this which causes the redshift. The further away a galaxy is, the longer the light has been travelling to get to us and so the more redshifted it will have become.

Chronometric cosmology is a contentious issue in astronomy at present. In fact, many astronomers seem to have decided that the best way to deal with it is to ignore it and carry on in their endeavours to prove the Hubble law beyond doubt. If the newer theory continues to predict values to much greater accuracy than the Hubble law, however, the tide of astronomical opinion will gradually turn and the whole basis upon which the big bang theory is based will be seriously undermined.

Whether the big bang is proved to have been right all along or whether one of the intrinsic redshift theories gains acceptance, cosmology is an exciting area of science at the present time. It is a subject which did not exist before Hubble and yet, in less than a century, it has gained principal importance in astronomy! It is a field which is so full of rich ideas that the open minded astronomer can wallow in this gold mine of hypothetical considerations and wonderful observations. Even if the big bang is eventually toppled, there can only be winners in this race, since our understanding of the universe will have advanced beyond measure. If the big bang is proven it will be a more robust theory than it might have been, thanks to the work of the intrinsic redshift astronomers forcing others to investigate more closely their preconceived ideas.

To reiterate a point which has cropped up a few times already, the wealth of new redshift data which will soon be flowing in from the latest survey instruments should turn the next decade into the most exciting yet for cosmologists. An ancient greeting says, "May you live in interesting times". If you are interested in cosmology, then you most certainly do!

Bibliography

Cosmology is a fascinating subject and I have only really scratched the surface in this book. I would recommend any of the following as excellent further reading.

A Brief History of Time *Stephen W. Hawking, Bantam Press, 1988 —* The bestseller of popular cosmology books!

Afterglow of Creation *Marcus Chown, Arrow, 1993 —* Tells the story of the scientific study of the microwave background from the point of view of the scientists involved.

A History of Astronomy from 1890 to the Present *David Leverington, Springer-Verlag, 1996 —* Excellent first point of reference for any historical perspective on astronomy.

A Short Course in General Relativity *J. Foster and J.D. Nightingale, Springer-Verlag, 1995*

A Short History of the Universe *Joseph Silk, Scientific American Library, 1994 —* Excellent popular level book with just a hint of mathematics.

Cosmology *Michael Rowan-Robinson, Oxford University Press, 1996*

Cosmology a first course *Marc Lachième-Rey, Cambridge University Press, 1995*

Dictionary of Astronomy *Jacqueline Mitton, Penguin, 1993*

Dictionary of Physics *edited by Valerie Illingworth, Penguin, 1990*

General Relativity, a first course for physicists *J.L Martin, Prentice Hall, 1996 —* A mathematical introduction io Einstein's work.

In The Beginning *John Gribbin, Penguin, 1993 —* Interesting, popular level book with some provocative ideas about the living universe.

Journeys to the Ends of the Universe *Chris Kitchin, Adam Hilger, 1990 —* Well written essays on some of the most difficult concepts in modern astronomy and cosmology.

Optics *Eugene Hecht and Alfred Zajac Addison-Wesley Publishing Company, 1974 —* Contains an excellent brief history on optics.

Principles of Cosmology and Gravitation *Michael Berry, Adam Hilger, 1989*

Quasars, Redshifts and Controversies *Halton Arp, Interstellar Media, 1987 –*
The personal account of one astronomer's work which has led him to believe
that the redshift is not solely a measurement of distance. Should be read by
everyone before they subscribe to the big bang party line, as some of the
evidence in here is hard to dismiss.

Towards the Edge of the Universe *Stuart Clark, Wiley-Praxis, 1997 –*
A review of the latest thinking in cosmology.

The Big Bang *Joseph Silk, W.H. Freeman, 1989*

The Farthest Things in the Universe *Jay M. Pasachoff, Hyron Spinrad,
Patrick S. Osmer and Edward Cheng, Cambridge University Press, 1994 –*
Four good, introductory essays on aspects of cosmology.

The First Three Minutes *Stephen Wienberg, Flamingo, reprinted 1996 –*
Before A Brief History of Time was published, this was the classic.

The Stuff of the Universe *John Gribbin and Martin Rees, Penguin, 1995 –*
Another popular level book which also introduces a little on the anthropic
cosmological principle.

Was Einstein Right? *Clifford M Will, Oxford University Press, 1995 –*
Excellent, non–technical account of Einstein's theory of general relativity.

Glossary

Absorption spectrum – The pattern of dark lines, caused by atomic absorption, superimposed on a continuous spectrum.

Accretion disc – The disc of dust and gas which swirls around a black hole.

Active galaxy – Any galaxy which is emitting large quantities of non-thermal radiation.

Amplitude – The size of the disturbance being propagated by a transverse wave.

Arc minute – One sixtieth of a degree.

Arc second – One sixtieth of an arc minute.

Astronomical wilderness – The volume of space between redshifts 5 and 1,000 in which galaxy formation began.

Atom – The smallest piece of matter which can still retain a chemical identity.

Big bang – The theory which describes the early evolution of the universe.

Binary star – A star system in which two stars orbit their common centre of gravity.

Black hole – A volume of space in which the density of matter is so great that not even light can escape from its gravitational attraction. Inside a black hole the known laws of physics break down.

Blueshift – The increase in electromagnetic radiation's frequency resulting from a decrease in the relative distance between the source and observer.

Broad line region – The region of an active galaxy in which fast moving hydrogen clouds produce large Doppler broadened lines in emission spectra.

Celestial sphere – An imaginary globe which surrounds the Earth and contains all the celestial objects on its surface.

Cepheid variable star – A regularly pulsating star which can be used to gauge distance in the universe.

Charge coupled device – An electronic camera which is capable of recording almost every photon of light which strikes its detector.

Chronometric cosmology – A rival theory to that of the expanding universe which proposes that the universe is static.

Continuous spectrum – The spectrum produced by a hot object. It contains no absorption or emission lines.

Cosmic kinematics – A branch of modern cosmology which studies the motion of galaxies but makes no attempt to understand why the galaxies move.

Cosmic microwave background radiation – A constant flux of electromagnetic radiation which has been redshifted into the microwave region of the spectrum. The photons of cosmic microwave background radiation outnumber the matter particles by one thousand million to one.

Cosmological constant – A constant which Einstein introduced into his equations of general relativity because he was unaware of the expansion of the universe.

Cosmological distance – A distance so great that light from any object situated that far away would take a significant fraction of the universe's lifetime to reach Earth.

Cosmological redshift – A redshift caused by the expansion of the spacetime continuum between us and the object in question.

Dark matter – Any form of matter which exists in the universe in a non luminous form.

Degree – One three hundred and sixtieth of a circle.

Doppler boosting – A Doppler blueshift of radiation emitted by a jet from an active galaxy. Doppler boosting takes place because particles in the emitting jet are travelling towards us at relativistic velocities.

Doppler broadening – The broadening of spectral lines because the emitting or absorbing clouds are swirling around a massive object at high speed.

Doppler effect – The alteration in frequency of electromagnetic radiation due to relative motion between the source and observer.

Electromagnetic spectrum – The entire range of radiation which can be propagated by a disturbance in the electromagnetic field of the universe.

Electromagnetic theory – The theory which describes one of the four fundamental forces of nature. It describes the electric and magnetic interaction between particles.

Electron – An negatively electrically charged, sub-atomic particle which is a constituent particle of an atom.

Electron degenerate matter – A type of highly compressed matter in which electrons are pushed very close the the atomic nuclei.

Emission spectrum – A spectrum which consists of nothing but coloured lines at very specific wavelengths.

Energy – A measure of something's ability to perform work.

Energy distribution – A measure of how much energy is being released at each wavelength in a spectrum.

Energy levels (around atoms) – Predicted by quantum theory, energy levels are the only orbits that electrons can exist in around atomic nuclei, unless unusual conditions apply.

Event – A happenstance in the spacetime continuum referenced by three spatial co-ordinates and a complementary temporal ordinate.

Event horizon – The boundary around a black hole at which communication with the outside universe ceases.

Expanding universe – A theory based upon the observation that distant galaxies display a redshift in their spectra.

Fingers of God – The name given to the elongation of galaxy clusters in maps caused because they are surveyed using redshifts rather than distances.

Fitzgerald-Lorentz contraction – A transformation which empirically describes why the Michelson-Morley experiment failed to detect the ether. Einstein later derived it from theoretical constraints in his special theory of relativity.

Frame of reference – A fundamental observing platform in the spacetime continuum.

Frequency – The measurement of the speed of a wave's oscillation.

Galaxy – A collection of matter which usually manifests itself by the production of stars.

Galaxy cluster – A collection of galaxies, bound together by the force of their mutual gravity.

Galaxy supercluster – A collection of galaxy clusters, bound together by the force of their mutual gravity.

Gaussian profile – The shape of a spectral line caused by the thermal oscillation of absorbing and emitting atoms.

General relativity – A theory of Albert Einstein's which describes gravity and can be applied to the universe on its largest scales.

Gravitational lens – A chance alignment between a nearby galaxy and a distant quasar. It results in the splitting of the quasar's appearance into two or more separate 'mirages'.

Gravitational well – The distortion in the spacetime continuum caused by the presence of an object with mass.

Gravity – One of the four fundamental forces of nature and the one most different from the other three.

Great Attractor – A gravitating mass, perhaps a cluster or supercluster of galaxies, which is pulling us and the surrounding galaxies towards it.

High mass star – Any star which is more than eight times the mass of the Sun.

High redshift galaxy – Any galaxy found to exist at a redshift of greater than 0.1.

Hubble constant – The constant of proportionality in the Hubble law. Its value must vary with time, so it is often referred to as the Hubble parameter. The Hubble constant is generally used to mean the value of the Hubble parameter at the current epoch and is somewhere between 50 and 100 km/s/Mpc with possibly a value close to 75 km/s/Mpc.

Hubble Deep Field – An image of one tiny part of the universe, taken by the Hubble space telescope, which shows thousands of galaxies at many different distances.

Hubble law – The linear proportionality, noticed by Hubble, between the distance of a galaxy and its redshift.

Hubble time – The time it would take for a galaxy to double its distance from the Milky Way. This means it can also be used as an estimate of the age of the universe.

Intrinsic redshift – A generic name for any redshift, which is not caused by any of the known methods, which is mistaken for a cosmological redshift and leads astronomers to over-estimate the distance scale of the universe.

K-Trumpler effect – An unexplained effect by which the spectral classification of stars, especially the blue/white stars, appears to influence their redshifts.

Lundmark law – A redshift-distance law which postulate a quadratic law between the two quantities rather than a linear law as used by modern astronomers.

Light – A restricted section of the electromagnetic spectrum to which our eyes are sensitive.

Local Group – A collection of several dozen galaxies to which the Milky Way belongs.

Local standard of rest – The average velocity of stars in the solar neighbourhood.

Longitudinal wave – A wave motion in which the disturbance occurs in the same direction as the wave's propagation.

Look-back time – The time it takes for light to reach us from a distant celestial object.

Low mass star – Any star which has a mass lower than eight times that of the Sun.

Mass – A fundamental property of matter in the universe. Mass defines an object's reluctance to change its state of motion and its effect on the spacetime continuum.

Medium – Any substance through which a wave motion can travel.

Messier Catalogue – A catalogue of just over 100 nebulae and galaxies, compiled by Charles Messier in the 1700s to help comet hunters recognise false alarms.

Michelson-Morley experiment - A famous experiment which failed to detect the ether.

Microlensing – A gravitational lens caused by the alignment of a star and a distant celestial object. Although not capable of creating multiple images, it brightens the distant, lensed object.

Milky Way – The galaxy in which we live.

Mössbauer effect – When certain atomic nuclei emit gamma rays, the nuclei recoil, removing some energy from the gamma ray and preventing it from being absorbed by another identical nucleus. This effect provides a very sensitive indicator of the change in a photon's energy and can be used to study the gravitational redshift of the Earth.

Narrow line region – The region of an active galaxy, usually surrounding the broad line region in which hydrogen clouds are moving more sedately and therefore only produce moderate Doppler broadened lines in emission spectra.

Nebula – A cloud of dust or gas in space.

New General Catalogue – A catalogue of 7,840 galaxies and nebulae, compiled by J.L.E. Dreyer and published in 1888.

Observer – Anything in receipt of electromagnetic radiation.

Pauli's exclusion principle – States that particles with half integer spins cannot occupy the same quantum states. This manifests itself as the reason why solid objects cannot exist in the same physical space.

Peculiar velocity – The motion of a celestial object relative to its local standard of rest.

Pencil beam survey – A deep photographic survey which maps a tiny region of space to a large redshift.

Photon – The fundamental quantum particle of electromagnetic radiation.

Planck constant – The fundamental constant of nature which defines the quantum regime.

Principle of equivalence – States that accelerating frames of reference are indistinguishable from the effects of gravity on frames of reference.

Proper motion – The apparent motion of a celestial object across the celestial sphere. Proper motion is caused by the combination of radial and transverse velocity.

Quantum theory – A theory which seeks to explain that the action of forces is a result of the exchange of sub-atomic particles.

Quasars – An intensely bright extragalactic object which superficially resembles a star. Most exist at very high redshifts and are therefore thought to be the nuclei of active galaxies.

Radial velocity – The component of a celestial object's velocity which is in a radial direction from the observer.

Radio astronomy – The study of the universe by the collection and analysis of radio waves.

Redshift – The decrease in frequency of electromagnetic radiation, brought about by an increase in the relative distance between the source and observer.

Redshift-distance relationship – The observed phenomenon whereby the more distant a galaxy is measured to be, the greater the redshift it displays.

Rest wavelength – The wavelength of an electromagnetic radiation emission which is observed if there is no relative movement between the source and the observer.

Schwarzschild radius – The radius at which a given mass turns into a black hole.

Sky plane – A tiny part of the celestial sphere which possesses a negligible radius of curvature.

Source – Anything which is emitting electromagnetic radiation.

Spacetime – A four dimensional framework in which events take place.

Special relativity – A theory derived by Albert Einstein which explained how observers in relative motion could compare their observations.

Spectral ratio – The ratio of electromagnetic wavelengths from different cosmic epochs. This gives the expansion factor of the universe.

Spectrometer – A device used to split the light from celestial objects into a spectrum.

Spectroscopic binary – A binary star which is only revealed to be binary by the movement of spectral lines in its spectrum.

Spectrum – The coloured pattern obtained by splitting light into its constituent wavelengths.

Spin – A quantum property of all particles which denotes the intrinsic angular momentum of the particle.

Spiral nebula – An archaic name for spiral galaxies.

Supernova – The explosion of a high mass star.

Thermal radiation – The characteristic release of radiation by a source which is at a temperature greater than that of its surroundings.

Thought experiments – Experiments which are performed in a physicist's head because they are impractical or undesirable to perform in practice.

Time dilation – The apparent reduction in the speed of time's passage observed in moving frames of reference.

Tired light – An unproven theory that light's passage through a static spacetime continuum can impart a redshift.

Transverse Doppler shift – The component of the special relativistic Doppler effect which is entirely due to time dilation.

Transverse velocity – The component of a celestial object's velocity which is at right angles to its radial direction from the observer.

Transverse wave – A wave in which the displacement occurs at right angles to the direction of the wave's propagation.

Tully Fisher method – A method of determining distance to galaxies based upon the galaxy's rotation. It does not rely on the galaxy's cosmological redshift.

Wavelength – The distance between wave crests or compressions.

Wavelength shift – The forced change in the wavelength of radiation.

Wave-particle duality – A principle of quantum theory which states that sub-atomic particles can behave like wave motions and vice-versa.

White dwarf – A small, highly compact stellar remnant.

Wolf effect – A physical process in which the emission of photons with very similar wavelengths can cause an overall wavelength shift in the observed radiation.

Zone of avoidance – The swathe cut by the Milky Way through the sky which makes the comprehensive observation of distant galaxies impossible.

Index